S0-AQM-738

ALLOSTERIC REGULATORY ENZYMES

ALLOSTERIC REGULATORY ENZYMES

by

Thomas Traut

Department of Biochemistry and Biophysics
University of North Carolina
Chapel Hill, NC, USA

 Springer

Thomas Traut
Department of Biochemistry and Biophysics
University of North Carolina
420 Mary Ellen Jones Bldg.
Chapel Hill, NC 27599-7260
USA

ISBN 978-0-387-72888-9   e-ISBN 978-0-387-72891-9

Library of Congress Control Number: 2007934262

Printed on acid-free paper.

9 8 7 6 5 4 3 2 1

springer.com

To Karyn Traut

Author's Note

This book has benefited from the careful scrutiny of my colleagues, who offered insightful comments, thoughtful suggestions, and editorial assistance. For their support I am grateful to Richard Wolfenden, Gottfried Schroeder, Randy Stockbridge, Charles Lewis, Jr., and Brian Callahan. The drawing in Figure 4.2 was provided by Jason Traut.

CONTENTS

SECTION 2. *K*-TYPE ENZYMES

SECTION 3. V-TYPE ENZYMES

SECTION 1

OVERVIEW OF ENZYMES

INTRODUCTION TO ENZYMES

Summary

All chemical reactions necessary for life are sufficiently slow that one or more unique enzyme catalysts are required to accelerate the reaction and make the needed product almost immediately available. Almost all enzymes are proteins that fold into domains. The majority of enzymes contains one domain (simple enzymes), while many are composed of two or more domains (allosteric enzymes and multifunctional proteins). Most enzymes are designed to function at a constant rate, but allosteric enzymes are sensitive to physiological controls, and thereby adjust their rate and determine the flux through the metabolic pathway that they control. There are two major groups of allosteric enzymes. One group is regulated by changing their affinity for one substrate, while keeping their maximum rate fairly constant (K-type enzymes). The second group also demonstrates significant changes in affinity, and in addition has large changes in the maximum rate (V-type enzymes). For cells to survive, natural selection has provided that each enzyme is always fast enough, with the slowest enzymes having a rate of $\geq 1 \text{ s}^{-1}$.

1.1 Introduction

All enzymes are remarkable for their ability to bind one or two substrates with appropriate specificity, and then facilitate a particular type of chemical reaction, producing one or more new products that are essential for the function of a living cell. Enzymes can be amazingly fast: for normal chemical reactions we have the example of a rate of greater than 10^6 s^{-1} for catalase,[1] and for carbonic anhydrase.[2, 3] Enzymes can perform exceedingly difficult reactions: for orotidine monophosphate (OMP) decarboxylase, the rate for the decarboxylation of OMP by the enzyme is 10^{17} faster than the spontaneous rate in the absence of enzyme.[4]

Over 5,000 different enzymes have been characterized, and almost all of these are proteins. If not stated otherwise, it will be assumed that any enzyme is a protein. A limited number of catalytic reactions have been demonstrated with certain types of RNA molecules, and such catalytic RNAs are now called ribozymes.[5, 6] These first two types of enzymes are normal biological molecules that have evolved to have the features that make them so essential. Based on the properties of these two types of normal catalysts, scientists have explored how to make novel catalysts with DNA and antibodies. The first such DNAzyme was designed to cleave RNA molecules,[7] but no natural DNAzyme has as yet been observed. A limited number of artificial enzymes have also been made by manipulating antibodies to favorably bind a reactive intermediate for some chemical reaction.[8] Such catalytic antibodies are also known as *abzymes*,* and are a demonstration of scientific ingenuity, even though these artificial catalysts are as yet very modest in their catalytic rates.

1.1.1 Why are Enzymes Needed?

Living cells have successfully evolved by adapting to two opposing needs. Their molecules should be stable under most conditions, yet the cell must be able to modify molecules or make new molecules as conditions require this. The organic molecules that have become the basis for cellular metabolism and life must be sufficiently stable to serve as structural units, information storage, catalytic agents and perform various other functions during the lifetime of any cell. These molecules are therefore maintained by bonds that are fairly stable, and such molecules commonly display remarkably long stabilities of many years in an aqueous solution, such as the cytoplasm of a cell. For example, the halftime of hydrolysis ($t_{1/2}$) in aqueous solution is about 400 years for proteins and about 140,000 years for DNA.[9] By comparison RNA has a $t_{1/2}$ of only 4 years.[9] Therefore, except when attacked by some reactive species, most biological molecules are quite stable in their normal cellular environment. At the same time, cells must be dynamic, with the ability to make new proteins and other molecules, and dispose of old ones continuously, in order to be successful in whatever environment they inhabit. The success of living organisms depends on this ability to have a stable cellular environment, as well as catalytic enzymes that can be controlled as to when and how they modify and manipulate all the molecules in the cell.

The stability of a molecule, or its thermodynamic energy, is illustrated in Fig. 1.1. It is the height of this energy barrier $\Delta G^{\ddagger}$ that defines the stability or the reactivity of a molecule. While chemical reactions may be enhanced in the presence of an acid or alkaline solution, or by a metal cation, enzymes have the unique ability to bind molecules with sufficient affinity to transiently stabilize their transition state (denoted by $S^{\ddagger}$ in Fig. 1.1), which greatly reduces the energy barrier, and thereby makes the transition between S and P vastly more favorable. The magnitude of this rate enhancement has been measured for various types of chemical reactions. The catalytic rate of an enzymatic reaction (k_{cat}) is generally at least a billion times greater than the nonenzymatic uncatalyzed reaction (k_{non}), and examples of the remarkable rate enhancement of various enzymes have been defined by Richard Wolfenden and colleagues, and are shown in Fig. 1.2. The most dramatic examples are illustrated by arginine decarboxylase (ADC)

---

*A contraction from ab (abbreviation for antibody) and enzyme.

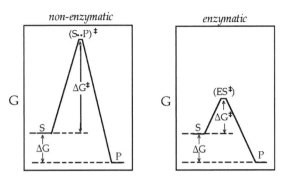

Fig. 1.1. An energy barrier ($\Delta G^{\ddagger}$) prevents the facile interconversion of S and P. An enzyme lowers this energy barrier by stabilizing the transition state between S and P, ES‡

and OMP decarboxylase (ODC). OMP decarboxylase is remarkable in that it has no cofactors to assist in this difficult reaction.[10] Orotidine-5′-monophosphate (OMP) is an intermediate in the biosynthesis of the pyrimidine nucleotide uridine-5′-monophosphate (UMP). OMP has a carboxyl group at carbon 6 of the pyrimidine base, and this must be removed to produce UMP. Without an enzyme to assist the decarboxylation, the elimination of this carboxyl group has a $t_{1/2}$ of 78 million years, demonstrating that this is a very stable bond.[4] The enzyme OMP decarboxylase performs this reaction about 25 times per second, providing a rate enhancement of 17 orders of magnitude.

An additional important point is also demonstrated by Fig. 1.2 with carbonic anhydrase (CAN). The hydration of carbon dioxide to form carbonic acid and bicarbonate is an extremely simple chemical reaction, and occurs with a $t_{1/2}$ of about 5 s in the absence of a catalyst. This spontaneous rate is still not fast enough for living organisms. The function of this enzyme is to hydrate carbon dioxide, a waste product of normal metabolism, and thereby produce carbonic acid, which spontaneously dissociates to bicarbonate, the major buffering agent in most organisms. Carbonic anhydrase performs this reaction in about 1 μs, and is therefore found in all organisms. Humans actually have 11 isozymes of carbonic anhydrase, expressed in our many different tissues.

1.1.2 Allosteric Enzymes

A simplified scheme for three metabolic pathways is illustrated in Fig. 1.3. Depending on various other factors, a specific cell will not need each of the three metabolic end products in equal amounts, at all times. It is therefore desirable to control how much of each of these products is actually made. This control function has evolved in the subset of enzymes known as allosteric regulatory enzymes.

A specific metabolic pathway, as shown in Fig. 1.3, normally includes 3–9 different enzymes in a sequential pathway dedicated to the synthesis of a single necessary molecule. Such a metabolic pathway may be viewed as a linear assembly line, in which each separate enzyme has a unique task in the sequential synthesis of the end product. The figure shows an example of a precursor compound, molecule A, which may be used for the synthesis of three different products, P, Q, and R. The cells' need for each of these

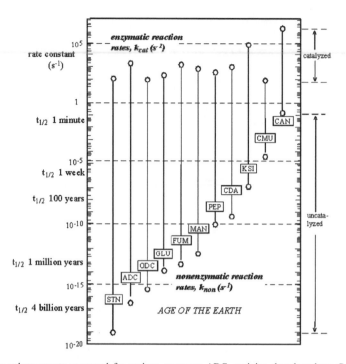

Fig. 1.2. Rate enhancements measured for various enzymes: ADC, arginine decarboxylase; CAN, carbonic anhydrase; CDA, cytidine deminase; CMU, chorismate mutase; FUM, fumarase; GLU, α-glucosidase; KSI, ketosteroid isomerase; MAN, mandelate racemase; ODC, OMP decarboxylase; PEP, carboxypeptidse B; STN, staphylococcal nuclease (figure courtesy of Richard Wolfenden)

final products may vary at different times, so that the three pathways have evolved to be independently regulated. The first enzyme that distinctly leads to that end product is normally the enzyme that commits the use of the substrate (B, in this figure) for the specific final product. Therefore, enzyme E_2 is the committed enzyme for the pathway leading to P. Enzyme E_2 is usually regulated by the end product, P. In this example, binding of compound P by E_2 would lead to this enzyme being inhibited, since this would occur only when P is at a high concentration, and its continued synthesis is no longer necessary. As the concentration of P becomes lower, since P is itself consumed over time, this inhibition diminishes, and the synthesis of P resumes. Such feedback inhibition by end products of the committed enzyme in a pathway is a standard feature in metabolism.

Enzymes that are able to be regulated by binding specific ligands are defined as allosteric (from the Greek: *allos* = other, and *stereos* = shape). This describes the key feature of such enzymes, their ability to change between two or more structural shapes that vary in their ability to bind a substrate, or in their ability to position a critical catalytic side chain, and therefore in their rate of catalysis. In Fig. 1.3, enzymes E_3 and E_4 would normally also be allosteric, but regulated by the end products Q and R.

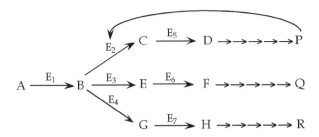

Fig. 1.3. A branched metabolic system. Enzyme E_2 is at the committed step for the synthesis of compound P. This end product normally acts as an allosteric inhibitor of the specific enzyme initiating the pathway for its specific synthesis

The typical examples presented show regulation by inhibition. It is common for these regulated allosteric enzymes to also respond in a positive fashion, with higher activity, to increased binding of a normal substrate, or an activator. It is by changing their rate, faster or slower, in response to changing concentrations of the specific cellular metabolites that these enzymes recognize and bind that enables such enzymes to be sensitive to some metabolic aspect of the cell. Since they respond by appropriately altering their activity, allosteric regulatory enzymes act as pacemakers for their pathway. We think of them as *regulatory*, since they regulate the pathway in which they function, and also because they are themselves regulated by the binding of physiological effectors.

Clearly, the needed feature for these pacemaker regulatory enzymes is that their activity or rate can be altered, and Nature has evolved two major strategies for regulating enzyme activity. Many enzymes are able to alter the affinity for their substrate, their K_m, with a conformational change. While such *K-type* enzymes have a fairly constant V_{max}, if at a fixed cellular concentration of substrate their affinity is made poorer, their rate must be slower, and as their affinity is improved, their rate will be faster. For the *V-type* enzymes the conformational change leads to a change in their affinity as well as in their maximum velocity. This may be accomplished by different means, such as the displacement, or the appropriate positioning of an important catalytic residue. Change in V_{max} may also occur by any factor that binds and sterically hinders access of the normal substrate to the catalytic site. Examples of these will be discussed in Chap. 1.3.

1.2 The Structures and Conformations of Proteins

1.2.1 Protein Conformations

A brief review of protein structure will help to explain enzyme binding sites, and the possibilities for allosteric effects. Most of the proteins in the cell, especially enzymes, normally fold so as to have an overall globular form. This comes naturally from the sequence of the protein, in which about one half of the amino acids are hydrophobic, and only when the protein folds so as to have these hydrophobic amino acids buried in the interior, away from the aqueous medium, will the form or structure of the protein be stable.

Each protein is a compact ensemble of secondary structures: the helices, beta strands, and loops that together comprise the total protein. While there is always an arrangement of these structural elements that is thermodynamically most favorable, variations from this most favored structure may not differ much in stability, so that most proteins are actually somewhat flexible, transiently converting into two or more somewhat similar structural shapes. Loops are especially mobile, and the opening and closing of loops at a catalytic binding site is frequently the rate-limiting feature for positioning of the substrate at the catalytic site. Binding sites are normally clefts or pockets in the surface of the protein, and may be formed by the proper positioning of adjacent secondary structure elements. Since proteins are flexible, the shape of the binding site may be transiently altered, as the overall shape of the protein varies. This feature provides the basis for regulation, by varying the fraction of the total enzyme population that has the correct shape or conformation to bind the desired substrate, and therefore the fraction of the total enzyme population that is competent to perform catalysis.

In all discussions about enzyme activity, and its regulation, it is important to think of each enzyme as a large population of molecules. Since enzymes generally have a cellular concentration above nanomolar, this denotes at least 10^9 enzyme molecules per microliter of cell volume for each specific enzyme. Never, under physiological conditions, will all of these molecules of the same enzyme have the same shape or conformation. The population will always include a mixture of several conformations or structural shapes, that is altered only by factors that may stabilize one of these conformational states, and thereby make it more abundant.[11,12] The illustration in Fig. 1.4, panel A shows the classical model of an allosteric enzyme that may have two conformations in the absence of a ligand, R and T. T is at a lower energy state and therefore the more stable and the more abundant form. For allosteric enzymes R represents the active form, while T is the less active or inactive form. In the absence of any ligands, T is normally the dominant species for *K*-type enzymes, while for *V*-type enzymes the dominant species may be either form, depending on the individual enzyme. The presence of a substrate, S, or an activator, A, will stabilize the R conformation, while an inhibitor, I, will stabilize the less active T conformation. Detailed examples of such allosteric features will be presented in later chapters.

Also, in Chap. 4 we will explore in greater detail the fact that all enzymes, whether allosteric or not, have multiple conformations. For the understanding of Fig. 1.4, the important point is that for normal enzymes there is only one active state under physiological conditions, and its abundance is not altered by any feature of the enzyme assay. Allosteric enzymes may often be represented by two conformations, since the key feature is the availability of regulatory effectors to bind to and stabilize the active or the inactive conformation. As the availability of the effectors changes, the distribution of the enzyme between the two principle conformations is changed, and this provides the basis for allosteric regulation.

Figure 1.4b illustrates the various equilibria[†] between these forms. In the absence of any ligands, the thermodynamic equilibrium favors conformation T, and therefore only a

---

[†]A true chemical equilibrium does not occur within cells, and a steady-state ratio of the two conformations is a more accurate description. The term *equilibrium* will be used since that is generally more convenient, in that it covers all simple chemical systems.

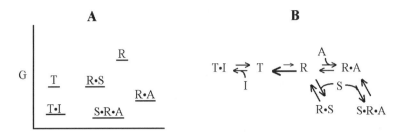

Fig. 1.4. Thermodynamic stability of enzyme conformations. G represents the free energy associated with any molecule. When proteins fold, they reach a stable tertiary structure that reflects their lowest free energy. T and R represent the inactive, and active forms of the enzyme. S, substrate; A, activator; I, inhibitor. The *dark arrow* in (**B**) emphasizes that allosteric enzymes will be proportionately more in the T form, since that is the more stable form. Note that ligands always stabilize (lower G) that conformation of the enzyme that binds the ligand

small fraction of the total enzyme population will be in the R conformation, which has better activity. Should the substrate become more abundant, it would bind to and stabilize the R conformation, making this species more abundant, and thereby increasing enzyme activity. This also demonstrates that the energy difference between these two conformations is very modest, since it cannot be greater than the binding energy of the substrate, which is normally in the range of 3–6 kcal/mol.

An activator that binds at a separate regulatory site would also increase the concentration of the active conformation. Overall, some of the enzyme molecules will always sample the different conformations, since energetically they are not that different. Depending on the enzyme, additional minor conformations may occur. Figure 1.4b is intended to illustrate the simplest system with only two conformations, though most proteins have more conformational states. However, if additional conformational states are not normally at a significant frequency, then the system may be simplified by considering only the conformations that are important for the observed enzyme activity.

It is important to note that the two conformations for active and inactive enzymes normally exist in the absence of regulatory effectors. The importance of such effectors is that they alter the equilibrium between the two conformations, and therefore alter the overall number of enzymes in the active conformation.

1.2.2 Protein Structures

The structure of a protein defines its function. A limited number of proteins form linear molecules, which serve as structural elements on a macromolecular scale. Silk and collagen are examples of such structural molecules that function in an extracellular environment, while myosin and fibroin are intracellular. Enzymes are almost always globular proteins, and they display a remarkable range of sizes, both for their subunits, and for the complete enzyme complex that many form. Protein structure is defined at four levels. The *primary* structure is the linear amino acid sequence of a protein; *secondary* structure defines the normal small structural elements such as alpha helices, beta strands, and loops; *tertiary* structure defines a single folded protein chain (equals a protein subunit); *quaternary* structure refers to the complex of two or more protein subunits.

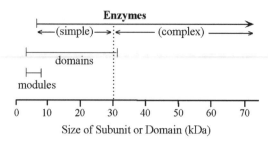

Fig. 1.5. Variation in the size of proteins, and in the size of structural components. The demarcation at 30 kDa is an approximation for single domain enzymes

Because of the large size range of proteins, illustrated in Fig. 1.5, additional terms have evolved to provide more specific descriptions about structure/function units within a protein. These are summarized in Table 1.1. It must be emphasized that currently there is no established consensus for the use of these terms. Different authors use these terms with somewhat distinct meanings, depending on what they wish to emphasize. In the following discussion, a protein's size or mass will always be for the single protein chain, or subunit, to avoid confusion with the size of large protein complexes. Enzymes that are proteins[‡] have sizes from as small as about 9 kDa for the HIV protease, and up to 565 kDa for the calcium channel in muscle cells.

In crystal structures of larger proteins (usually greater than 30 kDa), two or more distinct globular portions are frequently evident, and these are domains. Larger proteins always contain two or more domains. However, the term domain is also used to define a subcomponent of the protein by other criteria: the region that contains the catalytic site, or a portion of the protein that is easily cleaved by a protease, or the section of the protein that is involved in subunit contacts to form a dimer, and so forth. With an awareness of the different meanings associated with these terms, a reader can usually interpret the specific meaning by the context in which the term is used. At least half of all enzymes have a subunit mass of ≤ 30 kDa, and generally do not give evidence for containing more than

Table 1.1. Definitions for protein structural units

3° structure size	Term	Definition	M_r (kDa)
Large	Domain	Some subcomponent of total protein; "obviously" distinct	3–30
Small	Subdomain	Smaller local unit of 3° structure	3–20
	Module	Ligand binding unit Exon-coded unit	3–7
	Motif	An identified sequence associated with a specific structure/function	1.5–6

[‡]Proteins are extended chains of amino acids, and commonly when such chains are about 50 amino acids or less, they are defined as simple polypeptides, and begin to be called proteins as they become larger. There is no absolute size limit for the term protein.

one globular structural region. These are the simple enzymes shown in Fig. 1.5. Because the term domain have multiple definitions, domains overlap in size with simple enzymes, though they may occasionally be smaller. To help us with the discussion of protein evolution to follow, I will state that simple enzymes are one-domain proteins, and complex enzymes always contain two or more domains.

1.2.3 Multidomain Proteins

Ligand binding is one of the special functions of all enzyme domains, and when an enzyme has more than one domain, each domain commonly has a different ligand to bind. The term ligand (from the Latin *ligare* = to bind) includes all cellular metabolites that are substrates or effectors for enzymes, as well as macromolecules such as proteins, chromosomes, or membrane surfaces, to which enzymes may bind. It is a general feature that a protein's size is determined by how many ligand-binding sites it needs for its normal biological function. In other words, while enzymes may vary in size from 9 kDa to about 565 kDa, each enzyme is about the right size for its normal functions.

In the distribution of protein enzymes in simple bacteria we see that most of the enzymes are small. There is normally only one gene coding for each type of enzyme, but genes for enzymes that function together in a metabolic pathway are frequently clustered into an operon, a region of DNA that has the advantage that its genes are controlled by a single inducer region. When the gene for a catalytic subunit is adjacent to a second gene for a regulatory subunit that has the ability to bind to and alter the conformation of the catalytic subunit, then gene fusion can lead to these separate protein subunits becoming joined into a single protein subunit.

Gene fusion results when a termination signal at the end of the first gene is deleted or altered. Now, during transcription of this extended DNA segment the polymerase continues after the end of the reading frame for the first gene (no termination signal) and extends this RNA until it reaches the end of the second gene, producing a single mRNA that now codes for two domains, equivalent to the original two separate proteins. When the mRNA is translated, the two original proteins will no longer be separate proteins, but two domains that are joined by a short polypeptide chain encoded by the sequence of RNA between the two genes that was previously not transcribed or at least not translated. Naturally, to keep the second gene in the correct reading frame, the linker RNA region must contain $3n$ nucleotides. Also, the stop codon normally at the end of the first gene must be mutated to code for an amino acid, to assure continuity of the total polypeptide chain.

This simplest example would result in the formation of an allosteric regulatory enzyme, since it has combined the domain for a catalytic activity with the domain for binding regulatory effectors. By the same process, two or more genes for enzyme catalytic centers can become fused, if those genes are already sequential along a section of DNA.

1.2.3.1 Evolution of Multidomain Proteins

Gene duplication is a common event in most eukaryotes, and for living organisms in general it has been estimated that at least 50% of all genes were duplicated.[13] For humans,

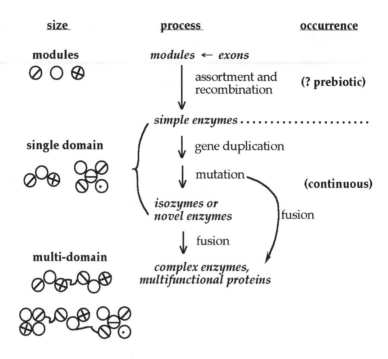

Fig. 1.6. Recombination of exon-coded modules, or larger DNA segments leads to various larger proteins

over 80% of our genes contain protein coding regions that are found in at least one other gene.[14] These extra copies of a gene (isogenes) may continue to code for essentially the same enzyme activity. However, since they initially are extra copies, then chance mutations may occur which modify the binding or catalytic rate of one of the duplicated enzymes, especially when this altered form of the same enzyme may also become preferentially expressed in a tissue or cell where the newer properties provide a benefit. Though the majority of such duplications are made nonfunctional by mutation, there are many examples of useful isozymes in mammals. It is this ability to benefit from mutations in extra genes that has led to new variants of the same catalytic function, or to important new enzyme activities.

An additional benefit of such duplications and recombinations is that these events may also be used in a new context, if they lead to the fusion of genes to produce proteins with two or more catalytic domains. These are known as multifunctional proteins, to emphasize that they contain more than one enzyme function. A simple scheme is illustrated in Fig. 1.6 to suggest how small protein modules, which normally are coded by a single exon, or larger domain-sized units may become fused to produce larger enzymes. In the majority of multifunctional proteins, the fused catalytic centers represent distinct different enzymes. There is frequently a ready comparison possible when the catalytic centers remain as individual enzymes in microbes, while having become fused into a single protein in higher eukaryotes.

One of the more dramatic examples of what is possible is given by the enzyme fatty acid synthetase. In bacteria eight genes code for eight different proteins,[15] of which seven have different catalytic activities that sequentially function to attach a two carbon acetyl group to a growing fatty acid chain, which is transiently anchored to the eighth member of this complex, the acyl carrier protein. Yeast give evidence that gene fusion has occurred, as yeast have only two genes, coding for five and three of the eight proteins for this metabolic sequence.[16] The fusion has become complete in mammals, as now a single gene codes for a single appropriately large protein containing all eight protein domains.[17] Since the seven catalytic activities of fatty acid synthetase work in concert on the acyl chain attached to the acyl carrier protein, but function sequentially, this organization of the eight distinct proteins in microbes into a single coherent protein in mammals represents greater efficiency.

The simplest system, and presumably the earliest version, is to produce eight proteins that function separately. Bacteria have partially improved on the simplest version, by having the separate proteins evolve sites to recognize and bind other members of this metabolic sequence, and then form a complex. Since all the enzyme subunits are required in comparable quantities for this entire assembly to be formed in a steady manner, fusion guarantees that each domain will be made in the same quantity as the others. By being linked into a single protein, each catalytic center is always present to optimize the steady and continued production of fatty acids.

Fatty acid synthetase is an example of the fusion of enzymes sequential in a pathway, and also arranged into a continuous group within an operon. There are also many examples of gene fusion of the same gene, when the second copy has occurred by gene duplication. Initially, this should only produce a protein with identical catalytic domains, for which there would be no clear benefit. But as with other isozymes, one of these two domains is now free to experience mutations, and if any beneficial mutation occurs this will then be selected. A spectrum of what is possible with the fusion of duplicated genes is in Table 1.2. With hexokinase II the duplicated domain retains the same activity, but one of the domains has an altered affinity for the same substrate, so that the duplicated enzyme now has a wider response range for the substrate than is normal for a single binding site.[18]

With carbamoyl-phosphate synthetase, the duplication of the carboxy kinase domain has led to modest changes so as to make it bind carbamate, which is formed from a product of the first domain, carboxy-phosphate.[19] With Hexokinase I the duplicated domain has lost catalytic activity, but now binds the product more tightly, and also still communicates with the active domain, so as to provide better inhibition by the product, glucose-6-phosphate.[20] Phosphofructokinase also uses the duplicated domain for several new regulatory sites to increase the number of effectors that control the rate of the active domain.[21] Finally, with the OPET decarboxylase/HHDD isomerase (Table 1.2) we have one example where sufficient changes were made in the duplicated domain to have it now perform a slightly altered type of chemical reaction. This is done without changing the normal ligand, since it isomerizes the product made by the first domain.[22]

For convenience, in the preceding discussion I have sometimes referred to changes occurring in the duplicated enzyme, or enzyme domain. When comparing such isozymes, we can only ascertain that they are highly related, and thereby derived from a common

Table 1.2. Possible new functions derived from gene duplication plus fusion

Added property	Enzymes	Function of new domain	Refs.
Same catalytic activity; extra substrate sensitivity	Hexokinase II	Same activity, but altered K_m for glucose	18
Same catalytic mechanism; different substrate	Carbamoyl-phosphate synthetase	A carboxy-kinase becomes a carbamate-kinase	19
New regulatory features	Hexokinase I	Inhibition by glucose-6-P	20
New catalytic activity	OPET decarboxylase/ HHDD isomerase[a]	New catalytic activity	22

[a]HHDD, 2-hydroxyhepta-2,4-diene-1,7-dioate; OPET, 5-oxopent-3-ene-1,2,5-tricarboxylate

ancestor. All the isogenes are equally susceptible to mutations. Once one of the isogenes develops beneficial mutations, it may replace the ancestor, or both may be altered and continue to function if their different attributes are beneficial and thereby maintained by natural selection.

1.2.3.2 Interaction Between Domains

The first successful model to describe cooperativity was presented in 1965 in a seminal paper by Monod, Wyman, and Changeux.[23] These authors introduced the simplest model, with two conformations for an enzyme, designated R, for the active conformation, and T, for the less active conformation. This model was quickly followed by a more extended treatment by Koshland, Némethy, and Filmer (KNF model)[24] who introduced the two visual depictions to represent the enzyme's conformation illustrated in Fig. 1.7, using a circle for T and a square for R. These visual shapes have now become icons for symbolizing the conformations of an enzyme, but they may be misleading in how readers visualize a protein structure. That is, the change from one conformation to the other suggests a complete alteration in the shape of the protein's subunit. When this model for allosteric regulation and cooperative conformational changes was presented in 1966, our knowledge of protein structures was still in its earliest phase, so that domains were not perceived as a general feature of protein structure.

In the ensuing years we have obtained thousands of protein structures, and a more advanced understanding of the details and variations that are possible. An important general finding is that conformational change in a protein is largely in the movement of

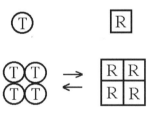

Fig. 1.7. The two classic conformational states introduced in the MWC model of Monod et al.[23] The visual icons for these conformations were introduced by Koshland et al.[24] Since allosteric enzymes are almost always oligomeric, an example of a tetramer is shown

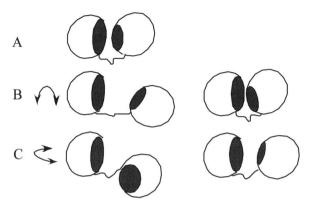

Fig. 1.8. Domain movement produces conformational change. In (**A**) the proper position of two domains forms the active binding site. Since the domains are connected by a short, but flexible polypeptide segment, then one domain may rotate within the plane of the figure, (**B**), to make the site too open or too closed. In (**C**) rotation perpendicular to the plane of (**B**) will again deform the proper shape of the binding site. Either domain may rotate without significantly changing its own conformation

one domain in a subunit relative to a connected domain (Fig. 1.8). This figure shows a normal binding pocket, such as a catalytic site, formed at the junction of two domains. To change the enzyme's ability to bind a ligand at this site, it needs only to be modified a little. This is most commonly done by rotation of one domain relative to its partner, since a short and sufficiently flexible polypeptide segment links them.

Since domains are comparable in size to entire subunits for smaller enzymes, a similar architecture in enzymes that are oligomeric is possible for the formation or disruption of a catalytic site. An example of this is illustrated in Fig. 1.9, where it is evident that the subunit by itself is not designed to have the binding pockets for its two substrates in an appropriate juxtaposition to facilitate catalysis. If these subunits normally join to form a dimer that is symmetric at the interface between the subunits, then now a complete catalytic pocket is formed across the dimer interface. Naturally, there will be two fully competent catalytic sites for the dimer, consistent with the almost universal stoichiometry of one catalytic site per subunit.

It is often possible to detect if enzymes have such binding sites that form between subunits, by exploring the activity of the enzyme as it dissociates from the oligomer to its subunits, as reviewed by Traut.[25] Important variables that influence enzyme dissociation include: enzyme concentration, ligand concentration, other cellular proteins, pH, and temperature. All these variables can be readily manipulated in vitro, but normally only the first two are physiological variables. The only constraints on how far the enzyme may be diluted come from the inherent activity of the enzyme, and the sensitivity of the enzyme assay. If very dilute enzyme can still form a small amount of product during the assay time, this must be enough product to be detectable, or an incorrect conclusion will be made as to whether the dissociated subunit is still active. Despite these constraints, more than 40 enzymes have now shown a change in activity as a function of their oligomeric state.[25] No single database has compiled all the enzymes known to show the feature illustrated in Fig. 1.9, but this feature is likely to be more widely used.

inactive   *active*

Fig. 1.9. Formation of a complete catalytic site between subunits. A single subunit has the binding sites for the two substrates, A and B. Since they are far apart, a chemical reaction between them is not possible. If the subunits form a symmetric dimer, then the two binding pockets will now be properly aligned

An alternative mechanism for changing the association of domains in the subunits of oligomers comes from the recently observed phenomenon that is now called 3D domain swapping.[26] It has now been observed in the crystal structures for more than 40 proteins that in the oligomeric assembly, domains do not remain together in the subunit that contains them. If the connection between domains within a subunit is sufficiently flexible, then domain A of one subunit can interact with domain B of the neighboring subunit, so that the functional unit is composed of domains from separate protein subunits. Seminal ribonuclease is an allosteric enzyme,[27] which also demonstrates domain swapping,[26] and this is postulated as the mechanism for its change in conformation.

This emphasizes the important concept about enzyme architecture, that in principle the interactions between domains in one subunit, and between subunits in one oligomer can produce equivalent structural features that are exploited for conformational changes, and make them possible. Because the domains, or the entire subunits, have very little change in their own conformation, the energy change involved in the conformations depicted in Figs. 1.8 and 1.9 are quite modest, and therefore make allosteric changes very favorable. This is then a general feature for understanding how proteins can be both stable and flexible, since the domains are normally fairly constant in shape, but any movement of one domain, by rotation relative to its partner, provides flexibility to adjust the shape of a binding site.

1.2.3.3 Alternate Oligomer Structures for the Same Enzyme

The classic view of enzyme oligomers is that a given enzyme will normally form only one type of oligomer, such as dimer, trimer, or tetramer. As depicted in Fig. 1.7, an enzyme may go through a conformational change, but is normally expected to maintain the same oligomeric assembly. An exception to this pattern has been demonstrated for human porphobilinogen synthase. The enzyme may form two types of dimer. The "hugging" dimer is stabilized by Mg^{2+}, and assembles to form the active octamer. However, in the absence of Mg^{2+} the octamer is unstable, and the detached dimer becomes more stable, and this dimer assembles to form a hexamer. These two oligomer forms have different pH optima and different affinities for the substrate.[28] A simplified scheme illustrates such subunit conformational changes (Fig. 1.10).

Crystal structures as well as appropriate kinetic studies define these two separate oligomeric structures for porphobilinogen synthase, and the morphological features for this enzyme have led to it being designated a *morpheein*.[28] Since somewhat comparable supporting data exist for a few other enzymes, it has been proposed that purine nucleoside phosphorylase, and ribonucleotide reductase are among additional enzymes that may be regulated in this fashion.[28]

1.3 Normal Values for Concentrations and Rates

In Chap. 2 we will consider enzyme kinetics, and it will be evident that many of these equations that define enzyme activity do not require a direct value for the concentration of the enzyme in any particular reaction. This format developed because it simplified the nature of the equations that were used to define kinetics, and it was considered acceptable because enzymes functioned as catalysts, and because the enzyme's concentration was not necessary for defining such features as the enzyme's binding affinity for its substrates. The accuracy of this assumption varies according to the particular metabolic step being defined, and also with the type of tissue or organism in which that enzyme functions. Depending on the cell's need for the product, the quantity of an enzyme varies accordingly, so that the quantity of product made per unit time is always sufficient for the metabolic needs of the cell. It is therefore instructive to consider actual values for both the concentrations of enzymes and of some major metabolites.

1.3.1 Concentrations of Enzymes

Scientists have not routinely measured the concentration of enzymes in cells, so no database exists for such information. However, we have enough information for some limited examples, and we can also make some educated deductions to provide approximate concentrations, which will help in our understanding of enzyme function.

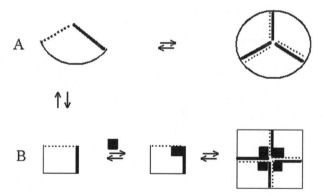

Fig. 1.10. Morpheeins may have alternate monomer conformations, which then assemble into different oligomers. Intersubunit contact requires joining a *solid line* to a *dotted line*. In the scheme shown, conformation A is more stable, and will therefore be more abundant and assemble to the trimer. If a special ligand (filled square) is present to stabilize conformation B, it will lead to formation of the tetramer

Cells generally contain at least 22–24% protein by weight relative to their volume.[29] This would then be as much as 240 g protein per L cell volume, and internally cells are about 70% aqueous.[30] We also know that an average protein has a subunit mass of about 30 kDa (= 30,000 g/mol). Then the calculated result for estimating the total protein concentration is about 11 mM:

$$\text{Protein concentration} = \frac{\left(\dfrac{240\,\text{g protein}}{30\,\text{g}/\text{mmol}}\right)}{0.7\,\text{L}} = \sim 11\,\text{mM}.$$

The human genome codes for just over 25,000 genes,[14] but these are not expressed all the time, or in all cells. With only 4,600 genes, *E. coli* produce all the enzymes for every metabolic pathway. With 6,300 genes yeast cells have the same metabolism, plus extra proteins to make a more complex cell membrane, and internal organelles such as a nucleus and mitochondria. It then seems plausible that a mammalian cell needs at most 10,000 proteins for normal metabolism, and any extra tissue-specific functions.

We may make the assumption that about 10,000 genes are expressed at any given time in a specific human cell. The remaining genes code for isozymes that are not expressed in the same cell. Therefore an average protein would be at a concentration of about 1 µM. Depending on the function of the protein, the concentration of individual proteins would largely range between 10 nM and 100 µM. We expect structural proteins (collagen, myosin, etc.) to be at the high end of the concentration range, and most enzymes to be average or below average in concentration. Due to the assumptions involved, these calculated values may have an error of 50% or greater, but they still give a useful range for discussing protein concentrations.

Individual enzymes should vary in concentration according to the cell's need for their product, and the enzyme's intrinsic rate. If the enzyme is slow, then more copies of that enzyme are required to produce the desired amount of product per unit of time. Compounds in central metabolism, related to energy production, protein synthesis, and nucleic acid synthesis, are at higher concentrations, because they are constantly used. Such metabolites are therefore more abundant than normal (Table 1.3), and we may expect enzymes in these pathways to be at higher concentrations. The ranges for each precursor type in this table are normally due to one or a few members having special functions. Except after a meal, most amino acids are normally below 1 mM. However, certain amino acids such as glutamate are more abundant, as they have additional roles. While all amino acids can be deaminated and used to form glucose or acetoacetate, glutamate is specifically required for energy metabolism since by deamination it can be converted to α-ketoglutarate, which functions as the most common acceptor for amine transfer from all the other amino acids, when these are being used for energy, instead of protein synthesis. In the brain glutamate also serves as an extracellular neurotransmitter.

Table 1.3. Concentrations of major metabolites

Important macromolecule	Normal precursor	Concentration (mM)
Proteins	Amino acids	0.1–3
Nucleic acids	Nucleotides	0.1–3
Carbohydrates	Simple sugars	0.1–2.5

Nucleoside-monophosphates and diphosphates are normally below 1 mM, and it is only ATP, which has the additional function as the most common phosphate donor in synthetic reactions, that reaches concentrations of 3 mM or even higher in some compartments.[31] In a standard metabolic chart, where each metabolite is represented by its name, and an arrow denotes each enzymatic reaction, different pathways appear visually to be very comparable. But a different visual metaphor may be more memorable for emphasizing that the opposite is true. Due to the constant need for energy, the volume of flow down the glycolytic pathway may be compared to the Mississippi river. It is simply the highest flux pathway in an average cell. Rates of RNA synthesis, and of protein synthesis, might compare to secondary rivers; these are still very large and active. Many other pathways might then be compared to smaller streams. With this sense of the varied flux for different metabolic "streams," it is then reasonable that the enzymes for glycolysis should be much more abundant than enzymes involved in other pathways.

Specific examples for the metabolites and enzymes of glycolysis are shown in Table 1.4. Both the concentrations of the metabolites, and the specific activities of the enzymes were measured in crude rat heart homogenates, without dilution.[32] While this made the measurements technically difficult and challenging, it permitted the authors to measure actual in vivo concentrations. I then calculated the concentration of each enzyme, by using the specific activity for the pure enzymes.[33] It is evident that all the glycolytic enzymes exist at concentrations above average for enzymes in general, as is expected for this pathway. It is also apparent that for four of these enzymes, their concentration is greater than the substrate that they bind, in clear distinction to the standard assumption that enzymes are at very low concentrations. We also see that enzymes that are present at high quantities are not always the slow ones, as might be expected.

This seeming anomaly may result from several different causes. First, the concentrations of the glycolytic metabolites are mostly quite low. The initial glucose is close to 2 mM, and the first phosphorylated compound, glucose-6-P is at about 170 µM. Most of the other metabolites are below 40 µM. Glucose-6-P is at a branch point, and in muscle this metabolite may be used for three distinct pathways, all of which may be active, but not at equal rates. There is reasonable evidence to support the hypothesis that the various enzymes are clustered into complexes in the cytoplasm and are not freely soluble, so that the metabolites of this pathway are efficiently catalyzed by the sequential enzymes, without the need for the intermediates to diffuse completely into the bulk cytoplasmic solvent.[34] This clustering of enzymes next to each other would provide an immediate benefit: intermediate metabolites could move directly from the catalytic center that makes them to the next enzyme in the glycolytic sequence.

Therefore, these metabolites would not diffuse extensively into the bulk cytoplasm, and such an arrangement would explain why the glycolytic intermediates are at such modest concentrations, when the pathway has such a high flux. To the extent that such enzyme complexes form, it would require that the members of the complex be present at concentrations comparable to each other. All would need to be present at a concentration similar to the slowest enzyme, whose abundance is dictated by the required overall rate.

Enzymes may have unexpected functions. An excellent example of organisms being opportunistic is the recruitment of normal enzymes for novel structural roles. Table 1.5 shows the example of the group of crystallin proteins, necessary to form the transparent

Table 1.4. Glycolytic enzymes and metabolites

Substrate	$[S]$ μM	Enzyme [a]	$[E]$ μM	k_{cat} (s^{-1})	M_r (kDa)
Glucose	1,910	Hexokinase II	3	192	100
Glucose-6-P	169	Phosphoglucose isomerase	20	511	59
Fructose-6-P	41	*Phosphofructokinase*	4	362	82
Fructose-1,6-P$_2$	35	Aldolase	53	1,980	54
Dihydroxyacetone-P	36	Triosephosphate isomerase	2	2,970	27
Glyceraldehyde-3-P	1.6	*Glycerald-3-P dehydrogenase*	34	156	36
1,3-P$_2$-glycerate	0.9	Phosphoglycerate kinase	712	353	46
3-P-glycerate	71	Phosphoglycerate mutase	65	174	29
2-P-glycerate	9	Enolase	20	94	47
P-enolpyruvate	13	*Pyruvate kinase*	15	637	57

[a]Enzymes in italics are allosterically regulated

structure of the lens of the eye.[35] It is evident that certain normal metabolic enzymes have the ability to form large-scale arrays that will align themselves with the required transparency for visible light to pass through. These enzymes would therefore be present at unusual concentrations in the lens. It is not clear if such altered functions permit these enzymes to maintain an enzymatic activity while in their cellular structures, but when isolated, some of these have been demonstrated to be enzymatically active in vitro.

Enzymes may be at higher concentrations because that is necessary to stabilize them. Enzymes are not equally stable in their physiological environment, and if they unfold this exposes them more to proteases, and leads to a shorter lifetime. It is thought that the ability of most enzymes to form oligomers helps to prevent this, as interactions between subunits help to stabilize their tertiary structure. Joining separate catalytic centers into a multifunctional protein may also increase such possible stabilizing of each catalytic domain. An example of this is provided by the comparison of the mammalian UMP synthase with its cognate enzymes in microbes. Orotate phosphoribosyltransferase (OPRT) and orotidine monophosphate decarboxylase (ODC) are distinct enzymes that catalyze the last two steps in the synthesis of the pyrimidine nucleotide uridine monophosphate (UMP). Via gene fusion, these two catalytic domains are linked into one protein in eukaryotes, UMP synthase. This fusion of two separate enzyme centers into one multifunctional protein has not significantly altered the intrinsic catalytic rates for the two domains of UMP synthase, relative to their microbial counterparts. One would therefore expect that microbes and mammals would need similar concentrations of these

Table 1.5. Proteins converted to crystallins[a]

Crystallin type	Organism	Homologous enzyme
α-crystallin	Mammals	Heat shock protein (HSP40)
β-crystallin	Mammals	Ca^{2+} binding protein S (bacteria)
γ-crystallin	Mammals	Ca^{2+} binding protein S (bacteria)
δ-crystallin	Crocodiles, birds	Argininosuccinate lyase
ε-crystallin	Crocodiles, birds	Lactate dehydrogenase
τ-crystallin	Fish, reptiles	Enolase
S$_{III}$-crystallin	Squid	Glutathione-S-transferase

[a]From Wistow and Piatigorsky[35]

Table 1.6. Concentrations of enzymes for UMP synthesis[a]

Cell/tissue	Concentration (nM)		
	[UMPS]	[OPRT]	[ODC]
Mammals:			
Human placenta	17		
Human lymphocytes	32		
Rat liver	11		
Rat brain	27		
Microorganisms:			
E. coli			2,900
E. coli		2,720	
Yeast		950	260
Yeast			375

[a]UMPS, UMP synthase; OPRT, orotate phosphoribosyltransferase; ODC, OMP decarboxylase. From Yablonski et al[36]

enzymes. But, the enzymes are always present in bacteria and yeast at concentrations far greater than for UMP synthase in mammals (Table 1.6).

One might suggest that much of this difference reflects the facts that the bacterial and yeast cells are grown in cultures, and therefore are largely in a cell mitotic mode, which needs transiently higher concentrations of these enzymes to supply the cells with nucleotides for DNA synthesis. But, for the mammalian examples, the lymphocytes were cultured cells, yet show only a modest increase in these enzyme concentrations.

Studies have demonstrated that the bifunctional mammalian UMP synthase is stable even when the enzyme is diluted to 0.1 nM, while the individual OPRT and ODC enzymes become unstable at 40 nM.[36] These results help to explain why microbial cells have such high concentrations of these enzymes, since this favors their forming into dimers, and this optimizes their stability.

1.3.2 How Fast are Enzymes?

The benefit of allosteric regulation is that the rate of a specific individual enzyme may be controlled, and thereby the flux of that specific metabolic pathway. But how fast should enzymes go? What rates would be too slow? Consider the encounter of an enzyme with one substrate, S:

$$E + S \underset{k_{off}}{\overset{k_{on}}{\rightleftharpoons}} ES \xrightarrow{k_{cat}} E + P \tag{1.1}$$

The catalytic rate for the formation of P, k_{cat} cannot exceed the encounter rate between E and S, k_{on}, which is proportional to the concentration of the two species, and the diffusion limit for the speed at which molecules move in an aqueous environment. We will explore this in more detail in Chap. 2, but I will simply note here that the upper limit for k_{cat} is about $10^7 \, s^{-1}$, and that a few enzymes approach that limit, with measured rates near $10^6 \, s^{-1}$ for catalase and carbonic anhydrase.[1,2]

Those enzymes currently shown to be very fast normally have the benefit of using only one substrate. These are hydrolases, isomerases, and mutases. Enzymes with a single substrate have the clear advantage that as soon as that substrate has bound, chemical catalysis may commence. Some examples of very fast enzymes are in Table 1.7. The majority of enzymes, using two substrates, must wait for the second substrate to bind before successful catalysis may occur. Actually hydrolases and hydrases bind water as the second substrate, but this will never be limiting as the concentration of water in cells is almost 40 M. Therefore, carbonic anhydrase is among the fastest enzymes.

It is more relevant to consider how fast enzymes must go for normal metabolic activities. Or, what is the slowest rate that is still fast enough? To put this in context, consider that one may purchase an automobile that can attain a speed of 200 mph. But if all of one's driving is on city streets, 25–40 mph is adequate, and for highway driving speeds of 65–70 mph suffice. And even if there were no legal speed limits, the extra speed possible with our super car is often not useful if the surrounding traffic is dense and moving normally. In cells, metabolic traffic moves at speeds that are satisfied by individual enzymes with rates generally between 10 and 1,000 s^{-1}. Eight of the enzymes in glycolysis (Table 1.4) have rates in the range 100–600 s^{-1}, and only two are faster than 1,000 s^{-1}.

While enzymes that are very fast amaze us, natural selection has ensured that the slowest metabolic enzymes have catalytic rates of ≥ 0.1 s^{-1}, since no natural enzymes have consistently been observed to be slower. And since only a few enzymes have been observed at this low rate, it is possible that these exceptions are due to some limitation in the assay by which they were measured, and that in vivo such enzymes might be somewhat faster. But how slowly can a reaction occur and still be beneficial to a cell? As an example, the spontaneous formation of carbonic acid from H_2O and CO_2 occurs in about 5 s, equivalent to a rate of 0.2 s^{-1} (Fig. 1.2). This must not be fast enough since we have carbonic anhydrase perform this reaction in a microsecond. It is a useful conclusion that enzymes generally have a catalytic rate of ≥ 1 s^{-1}.

A metabolic pathway cannot proceed faster than its slowest enzyme. Since they function as the pace maker for their metabolic pathway, regulatory enzymes normally have average catalytic rates when activated, but merely need to alter their own rate to be slower than, or equal to, any other enzyme that may be the limiting step. However, since it is the total flux at the regulatory step that must be controlled, fast enzymes may also be regulatory, if their concentration is appropriately lower so that their total rate is consistent for their function. The three regulatory enzymes in glycolysis (Table 1.4) have very average catalytic rates of about 200–600 s^{-1}.

Table 1.7. The fastest enzymes known

Enzyme	k_{cat} (s^{-1})	Reaction	Refs.
4-Oxalocrotonase tautomerase	2.8×10^6	2-Hydroxymuconate ⇋ 2-oxo-3-hexenedioate	37
Catalase	1×10^6	$2 H_2O_2 ⇋ H_2O + O_2$	1
Carbonic anhydrase	1×10^6	$CO_2 + H_2O ⇋ H_2CO_3$	2, 3
Ketosteroid isomerase	7×10^4	5-Androstene-3,17-dione ⇋ 4-androstene-3,17-dione	38

1.4 Brief History of Enzymes

In developing our understanding of allosteric enzymes, we are very fortunate to live at a time when so much significant research has been accomplished that very meaningful models may be devised to explain enzyme activity and enzyme regulation. Much useful information has been obtained after 1970, due to the advent of molecular biology techniques, and great improvements in our ability to obtain high resolution protein structures.

It may be interesting to briefly review the origins of our understanding of enzymes. Early in the nineteenth century attention to the process by which starch was converted to a simple sugar, or the process by which cane or beet sugar (sucrose) could be fermented to alcohol increased with the discovery that the polarimeter could measure some of the products by the change produced in the polarization of light. It was therefore possible to quantitate the formation of products, and thereby begin proper studies of the factors that influenced these systems.

Perhaps the first paper to describe a soluble extract with enzymatic activity was by Payen and Persoz (1833),[41] in which they described the preparation of *diastase* (from the Greek diastasis = separate), the name they gave to a soluble extract that cleaved starch. Anselme Payen, with a degree in chemistry from the École Polytechnique, had become the manager of a commercial business that prepared and sold various chemicals. Although remarkable, it is also quite understandable that, in this very first paper dealing with enzymes, the title of the paper referred to the "application to the industrial arts" for such enzyme preparations.

It is also interesting that the Greek name that they chose for this novel enzymatic preparation, diastasis, has an ending that for French pronunciation is normally converted to "ase." Since one of the next enzymatic preparations to be named, invertase, used this same suffix for ending the name of an enzyme, this may be why the "ase" suffix became standard for enzyme nomenclature in the twentieth century. By the second half of the nineteenth century different researchers had observed catalytic activities for trypsin and invertase, and also the fermentation of sucrose to produce alcohol and carbon dioxide. Most of this work was done with yeast, and considerable debate developed as to whether such an activity was solely a function of the intact cell, as proposed by Louis Pasteur and many others, or if the activity was due to some molecular component within the cell.

Since the term *ferment* had become popular in both French and German publications as a name for the active agent (originally in studies of fermentation), the proposal of the new name *enzyme* (from the Greek; en = in, zyme = yeast) by Kühne[39] was important in focusing attention on molecules within the cell. This was further strengthened by Buchner's paper,[40] in which he described very detailed and careful steps to disrupt yeast cells and remove and solubilize the activity necessary for sugar fermentation. While Buchner clearly established that a soluble extract had this activity, which he named *zymase*, he stated that he was not sure if this zymase activity belonged to the new class of enzymes. Buchner noted that all known enzymes (in 1897) were simple hydrolases, while his zymase catalyzed the much more complicated process by which sucrose is converted to alcohol and CO_2. At that time Buchner could not perceive that a single molecule could not have all the catalytic functions that we now know to be the enzymes of the glycolytic pathway plus alcohol dehydrogenase (Table 1.8).

Table 1.8. Important dates in the emergence of enzymology

Date	Event	Refs.
1833	Payen and Persoz describe a soluble extract from yeast that converts starch to glucose, and name it *diastase*	41
1876	Willi Kühne suggests the term *enzyme* for molecules that had been known as ferments	39
1894	Wilhelm Ostwald defines a *catalyst* as increasing a chemical reaction, while itself remaining unchanged	42
1897	Eduard Buchner demonstrates that conversion of sucrose to alcohol and CO_2 can be catalyzed by a *cell-free extract* of enzymes	40
1902	Adrian Brown determines that enzymes have a maximum activity (now defined as V_{max}), and that they are inhibited by their product	43
1902	Victor Henri defines the *enzyme-substrate complex*, and therefrom an equation for enzyme activity relative to the available substrate concentration	44
1913	Leonor Michaelis and Maud Menten propose the *use of buffers* for storage and assay of enzymes, develop the concept of *measuring initial activity* before product can begin to inhibit, and derive an *equation that included both the maximum activity and the affinity constant*	45
1926	James Sumner crystallizes urease and proves it to be a protein, in support of the proposal that all enzymes are proteins	46
1958	Daniel Koshland Jr. introduces the concept of *induced fit* to explain how correct binding influences catalysis	47
1962	Gerhart and Pardee report the first sigmoidal kinetics for aspartate carbamoyltransferase	48
1963	Monod, Changeux, and Jacob propose the concept of *allosterism*	49
1965	Monod, Wyman, and Changeux present a theory for the kinetics of allosteric enzymes	23
1966	Koshland, Némethy, and Filmer present an extended model for the kinetics of allosteric enzymes	24

Just before Buchner, Ostwald, a physical chemist, in 1894 proposed the definition for *catalyst* as an agent that accelerates a chemical reaction, while itself remaining unchanged.[42] In 1902 two important papers were published. Brown introduced the concept of a maximum activity for an enzyme, and noted that if the enzyme were presented with much higher substrate concentrations, it could not act any faster.[43] He also provided specific experiments demonstrating that glucose, a product of the invertase reaction, is a good inhibitor, while lactose is not. That same year Henri published a paper postulating a transient enzyme–substrate complex, and proposed the first equation for enzyme activity as a function of the substrate concentration, and included a term for V_{max}, though not for affinity.[44] While Brown had also calculated the rates for his kinetic experiments, he used an equation that was not directly dependent on [S], and had no terms for maximum activity or for affinity (Table 1.9).

Michaelis and Menten made several excellent proposals for enzymology in 1913.[45] The concept of pH had just been introduced a few years earlier, leading Michaelis to test the pH optimum for invertase, the enzyme which hydrolyzes sucrose to form glucose plus fructose. In this seminal paper, they proposed using buffered solutions for enzyme storage and enzyme assays. Up to this time papers on enzyme reactions often showed measurements at 30 or 60-min time periods, with the assay lasting up to 10 h in unbuffered solutions (e.g., Brown[43]). Michaelis and Menten introduced the concept of measuring the initial activity upon mixing enzyme with substrate, so that no product would have accumulated to inhibit the activity. Most importantly, they derived the

Table 1.9. Early equations for enzyme activity

Author (year)	Published equation	Interpreted version	Refs.
Brown (1902)	$k = \dfrac{1}{\Phi} \log \dfrac{1}{1-x}$	$v = \dfrac{1}{t} \log \dfrac{1}{\left(1 - \dfrac{[P]}{[S]_o}\right)}$	43
Henri (1902)	$\dfrac{dx}{dt} = \dfrac{K\Phi(a-x)}{1+m(a-x)+nx}$	$v = \dfrac{V_{max}[S]}{1+m[S]+n[P]}$	44
Michaelis & Menten (1913)	$v = C \cdot \Phi \dfrac{[S]}{[S]+k}$	$v = V_{max} \dfrac{[S]}{[S]+K_m}$	45

equation for the equilibrium of the enzyme and substrate with the ES complex, with terms for V_{max} and K_m that continue to be used today.

The lively controversy between the vitalists, who held that intact, living cells possessed a unique, vital organization that enabled activity, and the more progressive school favoring independent molecular agents, had not totally been resolved. Some of the debate had now shifted to the nature of the active agent; was it a protein or something other and more complex? The first definitive answer came with the paper by James Sumner describing the purification of the enzyme urease, and its subsequent crystallization to demonstrate its protein nature in 1926.[46] This was important support for the proposal that all enzymes were proteins. The scientific camp that supported vitalism, and that was opposed to molecules as the catalytic agents, was sufficiently strong early in the twentieth century, that Sumner, who would later receive the Nobel prize for his work, had to request permission to teach his new results at his own university, Cornell. This personal history is described very well by Tanford and Reynolds.[50]

Reviewing the many results with various enzymes in 1958, Koshland proposed the concept of induced fit to explain how enzymes can discriminate for the correct substrate, and explained how this would allow them to ignore smaller, but incorrect analogs.[47] The findings for several enzymes that their kinetics did not follow the normal hyperbolic curve that had been observed in earlier studies introduced the need to comprehend and model cooperative, allosteric kinetic behavior. Enzymes with the new sigmoidal form of kinetic plot included aspartate carbamoyltransferase (1962),[48] isocitrate dehydrogenase (1963),[51] and deoxythymidine kinase (1964).[52] In 1963 Monod and colleagues proposed the concept of *allosteric* enzymes with an allosteric site for effectors that are not homologous to the substrates.[49] In 1965 Monod, Wyman, and Changeux proposed their model for the kinetics of cooperativity,[23] which will be described in Chap. 5. The following year Koshland, Némethy, and Filmer extended this model to add additional modes for conformational change.[24]

Table 1.10. Databases that provide information for enzyme structure or kinetics

Information	Database	URL
Enzymes, general	BRENDA	http://www.brenda.uni-koeln.de
Protein structure:		
Classification	SCOP	http://scop.mrc-lmb.cam.ac.uk/scop
Crystal structures	Protein Data Bank	http://www.rcsb.org/pdb/searchlite.html
Protein domains	Dali	http://www.ebi.ac.uk/dali
	Toulouse	http://prodes.toulouse.inra.fr/prodom/current/ html/home.php
Protein sequence:	Swiss-Prot	http://us.expasy.org/sprot/sprot-top.html

1.5 Useful Resources

1.5.1 Websites

Various websites readily available on the Internet have databases that are supported by government or academic organizations, and therefore should continue to be available as resources. Table 1.10 lists some of these.

1.5.2 Reference Books

The following books provide extended coverage for particular topics relevant to the discussion of allosteric enzymes.

1.5.2.1 General Enzymology

Structure and Mechanism in Protein Science, by Alan Fersht, (1999). W. H. Freeman, New York. A good presentation of enzyme structure and protein folding, enzyme kinetics and mechanisms.

The Lock and Key Principle, The State of the Art — 100 Years On, edited by Jean-Paul Behr, (1994). John Wiley & Sons, New York. A presentation of key advances in recent years.

1.5.2.2 Allosteric Enzymes

Allosteric Enzymes, Kinetic Behaviour, by B. I. Kurganov, (1982). John Wiley & Sons, New York. An extensive documentation of many different allosteric enzymes.

Allosteric Enzymes, edited by Guy Hervé, (1989) CRC Press, Inc., Boca Raton, Florida. A detailed presentation of the eight best described enzymes in the 1980s.

1.5.2.3 Enzyme Kinetics

Enzyme Kinetics, Behavior and Analysis of Rapid Equilibrium and Steady-State Enzyme Systems, by Irwin H. Segel, (1975). John Wiley & Sons, New York. The most thorough treatment of enzyme kinetics. The book focuses more on the kinetics, and how to analyze and interpret them.

Fundamentals of Enzyme Kinetics, by Athel Cornish-Bowden, (2004) Portland Press, London. Up to date and very comprehensive. Valuable insights on the background and supporting principles for many standard equations.

1.5.2.4 Ligand Binding and Energetics

Binding and Linkage, Functional Chemistry of Biological Macromolecules, by Jeffries Wyman and Stanley J. Gill, (1990). University Science Books, Mill Valley, California. A complete treatment of ligand binding studies and their analysis.

Ligand-Receptor Energetics, by Irving M. Klotz, (1997). John Wiley & Sons, New York. A discussion of fundamental concepts and theories.

1.5.2.5 Enzyme Chemistry and Mechanisms

Catalysis in Chemistry and Enzymology, by William P. Jencks, (1969). Dover Publications, Inc., New York. A comprehensive treatment of a complex subject.

Enzymatic Reaction Mechanisms, by Christopher Walsh, (1979). An extensive analysis of this topic, with many details from representative enzymes.

1.5.2.6 Enzymes in Metabolism

Enzymes in Metabolic Pathways, by Milton H. Saier, Jr., (1987). An introduction to the topic of metabolic pathways and their control by key regulatory enzymes.

Understanding the Control of Metabolism, by David Fell, (1997). Portland Press Ltd. London. The first book to describe Metabolic Control Analysis, as an approach to understanding and quantitating metabolic flux, by understanding the network of interactions for any important enzyme.

1.5.2.7 History of Enzymology

Nature's Robots, A History of Proteins, by Charles Tanford and Jacqueline Reynolds, (2001). A very well written narrative on the history of enzymes/proteins by two authorities who contributed to the topic, and interacted with many of the participants.

1.5.2.8 Hemoglobin

Hemoglobin: Structure, Function, Evolution, and Pathology. by Richard E. Dickerson and Irving Geis, (1983). Benjamin/Cummings Publishing Co. Inc., Menlo Park,

California. A well illustrated presentation of this important molecule, which was the original protein shown to be allosteric.

Mechanisms of Cooperativity and Allosteric Regulation in Proteins, by Max Perutz, (1990). Cambridge University Press, Cambridge. An intimate account of the development of this field, and some of the work that helped to establish our understanding of positive cooperativity in ligand binding, by one of the original workers and one of the great thinkers in this field.

2

THE LIMITS FOR LIFE DEFINE THE LIMITS FOR ENZYMES

Summary

There are natural constraints that limit enzyme concentrations between 10 nM and 10 μM. For signaling switches k_{cat}'s are very low, at 10^{-2}–10^{-5} s^{-1}. For metabolic enzymes k_{cat}'s must be ≥ 1 s^{-1}, and are generally 10–3,000 s^{-1}. It then follows that for metabolic enzymes K_m values are generally limited to be between 1 μM and 1 mM. While increased K_m values would enable much faster k_{cat}'s, there is a clear need for enzymes to be sufficiently discriminating since so many have affinity constants below 100 μM.

2.1 Natural Constraints That are Limiting

The Michaelis–Menten equation expresses the rate of an enzymatic reaction, v, as a function of two other variables, the concentration of substrate, $[S]$, and the affinity of the enzyme for this particular substrate, K_m.

$$\frac{v}{v_{max}} = \frac{[S]}{K_m + [S]}. \tag{2.1}$$

But the maximum activity, V_{max},* itself is a function of the concentration of enzyme, $[E]$, so that we now have four variables that jointly define any catalytic rate:

---

*Some writers object to the use of V_{max}, since this term does not represent a true maximum limit, but simply an upper limit that may vary according to the experimental conditions. If readers are aware of this caveat, then the use of this term will make it easier to be consistent with a large body of enzyme literature.

$$v = \frac{k_{cat}[E]_o[S]}{K_m + [S]}. \tag{2.2}$$

When doing reactions in a test tube, with an assay volume between 100 µl and 1 ml, scientists routinely vary each of these over a considerable range. Since both k_{cat} and K_m are intrinsic properties of enzymes that have been subject to modification by evolutionary selection, let us first consider the limits to the concentration of an enzyme within a cell.

2.1.1 The Possible Concentration of Enzymes is Most Likely to be Limiting

Bacteria such as *E. coli* have very small cellular volumes, with an average of about 1 µm³, and a range that goes below 0.5 µm³.[53] And only about 70% of this is the aqueous cytoplasmic volume wherein most enzymes will be located.[54] A simple calculation will show that for any enzyme to be present as only a single molecule in the smallest of these cells, this would equal a concentration of 0.5 nM.

Calculations:

$\text{Volume}_{cytoplasm} = (0.7 \times 0.5 \ \mu m^3)(10^{-12} \ cc/\mu m^3)(10^{-2} \ L/cc) = 3.5 \times 10^{-15} \ L^{\dagger}$

$1 \ \text{molecule} = 1/(6.023 \times 10^{23} \ \text{molecules/mol}) = 1.66 \times 10^{-24} \ \text{mol}$

$1 \ \text{molecule/bacterial cell} = \dfrac{1.66 \times 10^{-24} \ \text{mol}}{3.5 \times 10^{-15} \ \text{L}} = 4.7 \times 10^{-10} \ \text{mol/L}$

This minimal quantity is clearly not likely, since whenever the single enzyme becomes damaged or inhibited, the cell would lose that activity completely. As few as 20 enzyme molecules would make this concentration equal to 10 nM, and therefore this value is a more realistic lower limit, and a value of 100 nM may be a more normal operational limit, since it still only stipulates about 200 enzyme molecules of a given type for an active bacterial cell. With the expanded cell volumes of mammalian cells, this same limited number of molecules will then equal a concentration in the low nanomolar range, consistent with the data in Table 1.6.

In Chap. 1, I described the average concentration of enzymes as being about 1 µM for mammalian cells. Table 1.4 shows that for glycolytic enzymes in mammalian cells, concentrations above 2 µM are standard. In a bacterial cell this would be about 4,000 molecules for each of these enzymes. And such a micromolar concentration range is also seen in Table 1.6 for the bacterial enzymes, ODCase and OPRTase, which are in pyrimidine biosynthesis. Since glycolytic enzymes should be at the high end of the concentration range, given their constant work load, then this may be the upper limit for

---

†A volume of 3.2×10^{-15} L has been directly measured for *E. coli* cells.[55]

Table 2.1. Natural sources for chemical damage

Source	Intracellular agent	Formation rate (s^{-1})	Protective enzyme	k_{cat} (s^{-1})	Refs.
Oxygen	Oxygen radical ($O_2^{\cdot}$)	10^4	Superoxide dismutase	1×10^4	55, 56
	Hydrogen peroxide	$\geq 10^4$	Catalase	1×10^6	1
Metabolism	Acidity	$\geq 10^2$	Carbonic anhydrase	1×10^6	3, 57
UV light	High energy photon	0.1	Photolyase	0.4	58

concentrations of enzymes in general.[‡] This gives a very definite range limit for the number of each enzyme catalyst that may exist in a cell, and life is possible only when these enzymes function at an adequate rate at these limited concentrations.

2.1.2 The Rate for Enzymatic Steps Must Be Faster Than Natural, but Undesired and Harmful, Reactions

We have a natural sense for time frames that define human actions in the larger physical world. Due to a general enthusiasm for sports, many people have an idea of what the fastest rate is for running, or cycling, or swimming. We are not as interested in lower limits, though culturally we have some awareness of this with such expressions as "Rome was not built in a day." We will again see a range of rates, depending on the actual task to be performed, and guided by the principle that each enzyme must be good enough. A critical starting point for this question is the normal rate for insult to a living cell by the various, ever present sources of chemical and radiation damage, even though these are normally at low levels. Examples include damage from oxygen radicals that form spontaneously, from various types of chemical damage, and from radiation damage which is largely due to ultraviolet rays. Such damage may occur to almost any molecule in the cell, but has long term results mainly when it involves DNA. Therefore, at least a subset of enzymes, with the responsibility of preventing such damaging agents, or of reversing their effects, must have reaction times that are faster than the natural rates for damage, examples of which are in Table 2.1.

We intuitively expect that a damaging agent should not be allowed to exist even for a few seconds, since it may cause too much harm in that time. In addition, if the source appears to be at a high level for the cell, as in the example of the cell bathed by sunlight, then the effective exposure to the source of damage is constant for up to 16 h, or many cell lifetimes. While a single celled organism could clearly survive by remaining in environments that suffered no *UV* exposure, such as deep ocean bottoms, much of life has evolved by being directly dependent on solar energy, or by benefiting indirectly. And we currently have many examples of enzymes that negate oxygen radicals, and repair damage to DNA. The enzymes in Table 2.1 demonstrate appropriately high kinetic rates in this regard, as detailed below.

[‡]It is possible to insert special plasmids, containing a unique gene, into *E. coli* so that the protein coded by this plasmid is expressed at an excessively high concentration, approaching 5 mM. This is an unphysiological, aberrant state for these cells, and should not be seen as contradicting the discussion for normal concentration ranges.

2.1.2.1 Oxygen Radicals

The earliest life actually formed in an anaerobic environment, but with the advent of cyanobacteria a limited oxygen atmosphere was produced by 2.5 billion years ago. Although oxygen led to a dramatic increase in the diversity of microbes and then multicellular eukaryotes, it also provided a new source of toxicity, in the form of oxygen radicals. The formation of $O_2^{\cdot-}$ in *E. coli* occurs at a rate of 5 μM s^{-1}.[55] For the actual volume of this cell, this concentration equals about 10,000 oxygen radicals per second. The observed k_{cat} for superoxide dismutase is also 10^4 s^{-1}. This enzyme cannot go much faster since it binds two molecules of the superoxide radical. However, to maximize the removal of superoxide, the enzyme exists at cellular concentrations of 10 μM.[59] This results in an overall very effective rate, so that the enzyme is able to reduce the steady-state concentration of the $O_2^{\cdot-}$ to only 10^{-10} M. To assist superoxide dismutase in maintaining a maximum activity, an additional enzyme, catalase, removes the peroxide produced by the dismutase, so that there will never be any significant product inhibition. Catalase itself is also very fast, with a k_{cat} of 10^6 s^{-1}.[1] For rapidly growing cells such as bacteria, this low level of oxygen toxicity is no longer harmful. For very long lived organisms such as humans, this amount of toxicity is seen as a significant factor for cumulative damage leading to senescence.[55]

2.1.2.2 Metabolic Acidity

The normal metabolism of carbohydrates and fats produces carbon dioxide, which is hydrated to form carbonic acid, and the carbonic acid dissociates to produce bicarbonate and H$^+$. This is a potential source of acidity, but organisms have evolved proton pumps to excrete the acid protons, and retain the bicarbonate to act as a buffering agent against other sources of acidity. Additional acidity comes from the formation of lactate under anaerobic conditions, as well as the frequent formation of many organic acids from the sulfur containing amino acids and phosphates. For human metabolism about 80 mmol/day of acids are produced.[57] Approximating this standard rate to a bacterial cell leads to a production of about 110 protons per second for a bacterial volume. Carbonic anhydrase catalyzes the very rapid hydration of carbon dioxide, which dissociates to provide the bicarbonate used in buffering against acids. The activity of carbonic anhydrase in providing bicarbonate easily compensates for the metabolic rate of fixed acid production.

2.1.2.3 Ultraviolet Radiation

Ultraviolet exposure is constant during daylight hours. In vivo experiments with *E. coli* by Aziz Sancar and colleagues have demonstrated the formation of pyrimidine dimers in DNA at a rate of 0.1 s^{-1}.[58] These authors also measured the concentration of photolyase at about 17 molecules/cell (about 10 nM), and a repair rate of 0.4 pyrimidine dimers per second. While this damage rate is naturally a function of the intensity of the UV light, it was observed that over a range of light intensity, this number of photolyase molecules always maintained cell survival.

This low rate of 0.4 s^{-1} is a misleading assessment of this enzyme's activity. That is, the enzyme cannot repair more damaged nucleotides than exist. Unlike other enzymes that have access to a steady concentration of substrate molecules, photolyase must search for the infrequent damaged site. It binds DNA sufficiently well that it spends most of its time sliding along the DNA double helix, until it encounters a damaged site. Based on experiments where the enzyme could be excited by rapid laser pulse, the reaction time for the photochemical repair is on the order of 10^{-12} s, which is remarkably rapid.

We can now extend this concept regarding lower limits on enzyme rates to enzymes in general. Any required chemical reaction must occur faster than the lifetime of a cell. But any specific microbe cannot be too leisurely in its reproductive time, since then other species with faster rates will come to dominate the available resources. The natural driving force from competition will result in reproductive cycles that are fast enough for a species to maintain itself. Bacteria are the ancestral cells, and under optimal conditions of nutrients and temperature, they can undergo cell division to produce two cells in about 20 min. Since nutrients are at an optimum, this means that the concentration of the substrate is not a limiting variable. However, any necessary chemical reaction must normally occur many times within a cell's lifetime, since cell division is a cumulative process in which individual enzymatic reactions, such as the synthesis of the nucleotides required for the duplicate DNA strands, must be performed many times by each enzyme.

Since an *E. coli* genome consists of 4.6 × 10^6 base pairs, then 9.2 × 10^6 nucleotides must be produced in at most one half of the cell life time, 600 s, so that the many other steps required for cell division may also occur. If the number of each enzyme molecule is at 100 per cell (50 nM), then each must have a rate of 153 s^{-1} under cellular conditions, meaning that their k_{cat} must be somewhat higher. Since these bacteria are at the same time making almost an equal quantity of RNA (mRNA, rRNA, tRNA), then rates of nucleotides synthesis must actually be about twice as fast. While there are some approximations in this argument, it helps to set some lower limits on the concentration of enzymes, and therefore on their minimum catalytic rates.

There is clearly some flexibility in the final rate necessary as a function of the concentration of that enzyme. Similar to the calculations at the beginning of this chapter, one may readily demonstrate that for the smallest bacterial cell a concentration of 2,000 molecules equals 1 μM. For the calculation above to provide adequate nucleotides, at this higher concentration of 1 μM these same enzymes could satisfy their function with a k_{cat} 20-fold slower, at about 30 s^{-1}. Since enzyme concentrations are almost never above 10 μM, then at this upper limit these same enzymes could be slower, with a rate of about 3 s^{-1}, and still accomplish the needed production of nucleotides within the desired time limit. We again approach the lower rate barrier of 1 s^{-1}. But, since the total protein concentration is itself limited, then only some enzymes can reach such a high concentration, and the majority will clearly need to be faster. This helps to set some lower limits on the concentration of enzymes, and therefore on their minimum catalytic rates.

I have described here logical reasons to account for the observed concentration of enzymes in a bacterial cell, and these values correspond very nicely with the observed values for most enzymes. The not surprising conclusion is that living organisms, responding to the pressures of natural selection, have generally reached a state where

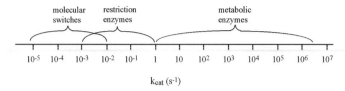

Fig. 2.1. The range for enzyme catalytic rates

their enzymes, as a total system, have reached an optimum balance between possible enzyme concentrations and the rates needed to maintain a dynamic and successfully reproducing organism. Since the majority of the estimated 20,000 enzymes in human cells have not yet been characterized, it is certainly possible that a few will emerge that do have k_{cat} values somewhat below 1 s^{-1}. A few such slower enzymes might be sustained by the system, if the greater majority remains consistent with the constraints that I have described.

2.1.3 DNA Modifying Enzymes: Accuracy is More Important Than Speed

I have shown various data to support a lower limit for enzymatic rates of about 1 s^{-1} for enzymes with normal metabolic functions (Fig. 2.1). There is a special group of enzymes whose function is to alter genomic DNA. They may methylate certain bases along the host's genomic DNA to transiently make such genes less available for transcription. They may also cut foreign DNA, belonging to invading viruses or other pathogens, into fragments to make it inactive. Such enzymes must be accurate as to where they modify the host DNA, and also be specific in recognizing restriction site sequences that are unique to the foreign DNA. This accuracy is achieved by greater slowness. Type III restriction enzymes have a k_{cat} of about 1 s^{-1},[60] while type II and type I enzymes have rates of $0.1–0.05 \text{ s}^{-1}$,[61, 62] and 0.001 s^{-1}.[63]

2.1.4 Signaling Systems: Why Very Slow Rates Can Be Good

There is a special group of enzymes where a much slower activity is necessary, since it defines a limited time for a signal to exist. These signals occur in processes where it is necessary to switch between states of activity, and to maintain the altered state for a transient, but defined period. Depending on the process, this transient time may last for only tens of seconds, for many hours, or even for many years. The defining limit for this transient period is the slow rate at which a key regulatory enzyme makes or cleaves a phosphate bond. Currently known examples include the various G proteins, enzymes that control circadian clocks, and enzymes involved in memory storage. These enzymes are also referred to as *molecular switches* (Table 2.2).

G proteins are themselves regulatory, having two different conformations when they are binding GTP or GDP. Since GTP acts as a regulatory signal, it stabilizes a new conformation in the GTP binding domain, which in turn influences the activity of an adjoining domain in the same multifunctional protein, or a separate target enzyme. The GTP binding domain has a very slow rate for the hydrolysis of GTP, which permits the

Table 2.2. Molecular switches – the slowest enzymes known

Enzyme	k_{cat} (s^{-1})	Reaction	Refs.
GTPase	1×10^{-2}	$GTP + H_2O \rightarrow GDP + P_i$	64
KaiC	10^{-3}–10^{-4}	$KaiC + ATP \rightarrow KaiC\text{-}P_{1\text{-}3} + ADP$	65
KaiC phosphatase	10^{-3}–10^{-4}	$KaiC\text{-}P_{1\text{-}3} + H_2O \rightarrow KaiC + 1\text{–}3\ P_i$	65
CaMKII	8×10^{-6}	$CaMKII + ATP \rightarrow CaMKII\text{-}P + ADP$	66

active conformation of this domain to maintain the regulatory stimulus on its neighbor for as long as the GTP remains intact. Upon hydrolysis of the GTP to form GDP, which occurs very slowly at a rate of about 10^{-2} s^{-1},[64] a new conformation occurs, which now has little influence on the neighboring catalytic function that it is intended to control. This slow rate of hydrolysis therefore serves as a built in clock that limits the duration of the regulatory signal to about 100 s. In addition, the GTP binding domain has tighter binding for GDP, so that this product is released slowly, so that the less active/inactive form of the enzyme is now stable for many minutes.

Cyanobacteria have a circadian clock that depends on the phosphorylation state of the protein KaiC.[65] KaiC acts to regulate gene expression in a circadian pattern. It has autophosphorylation and autodephosphorylation activities, and these two activities are regulated by the additional proteins KaiA and KaiB. KaiA stimulates the auto-phosphorylation, while KaiB attenuates this function. Both the phosphorylation and the dephosphorylation rates are remarkably slow (Table 2.2), so that it takes many hours to phosphorylate the protein, and a similar length of time to dephosphorylate. The alterations between these two very slow rates set the circadian pattern as the KaiC protein is converted to the phospho-enzyme state, and then to the native state.

A similar switch pattern is observed for CaMKII, a calcium/calmodulin dependent protein kinase that is involved in memory storage.[66] A memory impulse activates this enzyme by the release of calcium/calmodulin which bind to the hexameric enzyme, and induce it to begin autophosphorylation of that subunit, until the hexamer is completely phosphorylated and activated. The phosphorylated CaMKII can in turn be dephos-phorylated by a specific protein phosphatase. The duration of the signal is enhanced by the fact that the postsynaptic density contains only about 30 enzyme molecules.[67] The postsynaptic density is the visible structural region on the postsynaptic membrane that contains a highly structured complex of molecules.

2.1.5 What is the Meaning of the Many Metabolic Enzymes for Which Slow Rates Have Been Published?

In the literature over the past 50 years there are many published values for metabolic enzyme activities that are well below 1 s^{-1}. This is easily observed with a general data base, such as BRENDA.[33] Inspection of the published values for many enzymes often shows a range in the specific activity for the same enzyme of 100-fold or greater. I tend to trust the higher values. Unless one makes a significant error in recording the activity rate, or in its calculation, one cannot make an enzyme go faster than is normal for it. However, enzymes are often sensitive, and kinetics are done with enzymes that are not in their normal milieu. It is therefore not unusual that researchers observe low rates, since the enzyme may have become partly denatured during the purification procedure, or some aspect of the assay conditions are not optimal.

Among the most common problems are that intracellular enzymes function in a reducing environment, and those that have surface cysteines may form unwanted intra- or inter-subunit disulfide bonds in an oxidizing storage or assay buffer. Adding reducing reagents, such as dithiothreitol is now normally tested early in a purification. A better choice of buffer is sometimes needed. Phosphate makes an excellent buffer and is very economic. But, when assaying enzymes that bind nucleotides, the phosphate of the buffer will always be a background inhibitor that prevents measurement of the true V_{max}. Cells have many types of proteases that are often constrained in a special organelle (Golgi and endoplasmic reticulum). Disruption of cells to obtain the desired enzyme normally breaks these organelles, so that their proteases now have contact with the desired enzyme. Inhibitors of such proteases are now routinely employed in the early stages of enzyme purification. Further problems emerge with enzyme storage, or loss of a cofactor during dialysis, and so forth. The list of potential problems that are generally preventable can be daunting to new researchers. Because of the ease with which enzyme activity may be unwittingly decreased by the experimenter, caution and judgment are necessary in accepting some of the published rates for enzymes.

2.2 Parameters for Binding Constants

A few simple examples will help to clarify binding constants. To emphasize the general nature of this discussion, let us consider the binding of a proton by acetate, as shown in a normal titration curve (Fig. 2.2). Although the affinity of acetate for binding a proton is poor, since the pK_a is 4.8, it serves as a useful model. This binding constant, the pK_a, defines the concentration of the ligand to be bound, H^+ that is needed for 50% binding. For an approximately tenfold change in concentration above this pK_a, at pH = 3.8, the curve continues to be almost linear before reaching a plateau at 100% saturation. In the same way, down to a proton concentration tenfold lower than the pK_a, at a pH of 5.8, the curve continues to be almost linear before reaching a plateau where there is no binding. Since titration curves are always shown on log plots, it is then a simple mnemonic to remember that the effective range for binding is over almost 2 logs of the concentration of the ligand. This will be true for any binding interaction which occurs at a *constant affinity* by the receptor for the ligand being bound. What is demonstrated in Fig. 2.2 for the binding of a very small ligand, H^+, to a very small receptor, acetate, will also hold true for the binding of much larger ligands to normal enzymes.

2.2.1 The Importance of Being Good Enough

We know that enzymes should evolve to have a binding constant appropriate for optimizing their normal activity. But what defines normal activity for different enzymes? The two obvious constraints are speed and accuracy. If we consider three professions, neurosurgeon, barber, and candy vendor, we intuitively appreciate that we cannot expect

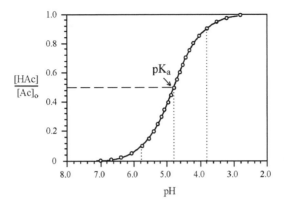

Fig. 2.2. Proton binding by acetate

of each one an equal number of transactions with patients/customers per day. Surgeons need to be very discriminating in what/where they cut. Their speed should be no faster than that speed at which they will make no error. The art of cutting hair is not quite as exact, and barbers can proceed at a moderate speed. Vendors may clearly proceed at faster rates, since they may safely correct occasional errors with no harm to the customer. In the spirit of this metaphor, we expect enzymes involved in DNA synthesis to be more stringent in binding the correct nucleotide to avoid mutations. The main requirement is that their error rate should be low enough so that a sufficient majority of organisms succeed in producing offspring without many mutations. Since the degree of fidelity in mammalian DNA synthesis has an error rate of $<10^{-8}$, it would not be effective to have an enzyme bind with such stringent affinity so as to accomplish this, for the catalytic rate would then be far too slow. An ingenious proof reading function has evolved, which divides the recognition of the correct nucleotide into two steps, so that neither has to be too stringent, and thereby limit the rate of DNA synthesis.

We have already seen the demand for speed in enzymes such as catalase and superoxide dismutase (Table 2.1). These enzymes work in sequential steps to neutralize oxygen radicals. These compounds are formed spontaneously in an oxygen environment, and are very damaging to DNA and therefore mutagenic. It is then not surprising that speed has been selected in enzymes that prevent oxidative damage (superoxide dismutase, catalase) or constantly replenish our buffering capacity (carbonic anhydrase). These enzymes have rates of about $10^{4}-10^{6}$ s^{-1}. When comparing enzymes, such as the glycolytic enzymes shown in Table 1.4, we see that triose phosphate isomerase is 30 times faster than enolase, and we may be suitably impressed by this very fast enzyme. But, it is also evident in Fig. 1.2 that not all chemical reactions are equally difficult. Therefore, although carbonic anhydrase is a thousand times faster than staphylococcal nuclease, it is the rate enhancement performed by staphylococcal nuclease that is truly astounding. While we will see a spectrum of values for both K_m and k_{cat}, as a general rule each enzyme has been selected to be at least good enough for its specific function.

Table 2.3. Estimated values of K_d consistent with a normal k_{cat}

k_{off} (s^{-1})	K_d (M)	k_{cat} (s^{-1})	
10^8	10^{-1}	10^7	
10^7	10^{-2}	10^6	normal limits
10^6	10^{-3}	10^5	
10^5	10^{-4}	10^4	
10^4	10^{-5}	10^3	normal range
10^3	10^{-6}	10^2	
10^2	10^{-7}	10^1	
10^1	10^{-8}	1	
1	10^{-9}	0.1	

2.2.2 The Range of Binding Constants

It has been observed that glycolytic enzymes have a K_m for their normal substrate that is equal, or at least comparable to the normal concentration of that substrate.[68] This feature permits some variation in the enzymatic rate as the concentration of its substrate varies under cellular conditions. Under normal conditions the enzyme would be 50% active if $K_m = [S]$ cell, and the enzyme would still have an almost linear response to the substrate, even if it declined or increased by about tenfold. But, would it be inefficient to have a binding constant significantly different from the normal $[S]$? Table 1.3 lists a few normal metabolites and their cellular concentrations. We might expect enzymes using such metabolites to have affinities comparable to these concentrations. However, it would be equally correct to say that cells arrange to maintain their metabolites at concentrations that are consistent with the K_ms of the respective enzyme. This may be the more meaningful constraint, since enzymes under selective pressure may evolve to have a binding constant that is good enough. To keep this from being a circular argument, let us first consider the limits for enzyme binding constants.

The on/off binding of the substrate is shown in (2.1). The on rate is assumed to be fairly standard for the encounter of enzyme and substrate, and the actual rate has been calculated to be as high as 7×10^9 s^{-1} M^{-1} for two molecules of equal size in water. In a more physiological medium of appropriate ionic strength, rates of 10^9 s^{-1} M^{-1} have been observed. Also, for the majority of enzymes k_{cat} is slower than k_{off}. Since k_{off} then determines the binding affinity, we may easily approximate the limits for both k_{off} and thence K_d using (2.3).

$$K_d = \frac{k_{off}}{k_{on}}. \tag{2.3}$$

Table 2.3 shows the calculated results for assuming $k_{on} = 10^9$ s^{-1} M^{-1}, when k_{off} varies between 1 and 10^8 s^{-1}. The values for K_d in this table are calculated with (2.4).

$$K_d = \frac{k_{off}(\text{s}^{-1})}{10^9 \,\text{M}^{-1}\text{s}^{-1}}. \tag{2.4}$$

To estimate values of k_{cat}, also assume that k_{cat} will be tenfold lower than k_{off}, though the true difference is frequently much greater. From these calculations we then see that in order to have the minimal activity of 1 s^{-1}, K_d should be no lower than 10^{-8} M. To have the highest activities so far measured, K_d can be as high as 10^{-2} M. This range of K_d values has been observed for different enzymes, and is close to the limit of what appears to be possible. These values define the boundaries for normal enzymes, although for most enzymes k_{cat} has values of 10–3,000 s^{-1}. Though we normally have a sense that faster enzymes should be better, only about 20 different enzymes have been shown to have kinetic rates greater than 10,000 s^{-1}.[33] Even for the enzymes of glycolysis, the highest flux pathway in the cell, no enzyme has a rate greater than 3,000 s^{-1} (Table 1.4).

A note of caution is necessary, since the assumptions used to calculate Table 2.3 do not absolutely apply to every enzyme. But, they are a useful guide for the majority of enzymes, and the values so calculated are very consistent with measured values that are currently known. These calculations then tell us that for enzymes to have normal rates, with a k_{cat} of 10–3,000 s^{-1}, they should have affinity constants of 10^{-7}–10^{-4} M. And these affinity values are quite comparable to the normal concentrations of the respective substrates.

Why are not affinity constants much lower than $[S]_{cell}$? Tight binding might be better, since it will give the best discrimination for the specific substrate. In accord with this hypothesis is the fact that there is a distinct, unique enzyme for almost every single reaction. As a simple example, purine and pyrimidine nucleosides need to be phosphorylated three times to produce the nucleoside-triphosphates that are essential metabolites. It might be possible to have a single kinase able to perform each of these reactions, if it had a nondiscriminating catalytic site at which any of the intermediates could bind. We find that there is almost a separate kinase for each nucleoside, and nucleoside monophosphate, with only a few enzymes serving two substrates. Only nucleoside-diphosphate kinase is able to bind and phosphorylate all of the nucleoside diphosphates. This demonstrates that the ability to specifically control each of the pathways leading to ATP, GTP, CTP, and UTP is sufficiently necessary that almost all organisms make the appropriately distinct enzymes for each step. Clearly, for an enzyme to be distinct, it must bind its specific substrate well, while binding close analogs fairly poorly. This means the binding constant for the normal substrate must be below 1 mM, and preferably below 100 µM. In Fig. 2.3, we see that three fourths of the K_m values are below 100 µM. But again, binding should only be as stringent as necessary, while not impeding the required catalytic rate. Therefore, only one K_m value is below 1 µM.

Why should not affinity constants be sufficiently higher than $[S]_{cell}$? A high K_m means poor binding, and that in turn means the rate can be much faster. If the enzyme does not bind the substrate tightly, it will not bind the product tightly, since most of the

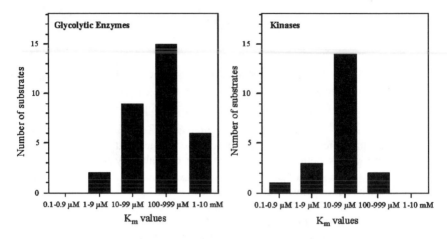

Fig. 2.3. K_m values for substrates of the glycolytic enzymes, adapted from Fersht,[68] and of nucleoside/nucleotide kinases. For the kinases the K_m values are for the acceptor substrate[69–86]

same binding determinants are in both of these molecules. If the product is not bound tightly, then it will normally dissociate very rapidly from the enzyme, so that the enzyme is again free to bind another substrate molecule and continue to be productive. But poor binding means that the substrate-binding site is not well defined, and similar molecules that resemble the substrate may also bind there. This means that the enzyme is no longer very discriminating. Sometimes this feature is desirable, as with general proteases which function in the catabolism of proteins to recover the amino acids.

More generally this poor discrimination may not be useful, and the need to control the synthesis or catabolism of most molecules has led to enzymes somewhat more specific for a preferred substrate, as suggested by the normal range and limits indicated in Table 2.3. If speed were sufficiently desirable, then by now we should see that need expressed in a lot of weak binding constants. Figure 2.3 shows such values for the ten glycolytic enzymes, as well as for a group of nucleoside and nucleotide kinases. For the glycolytic enzymes about three-fourths of the K_m values are below 500 µM, and over one-third are below 100 µM. On average, K_m values for the kinases are more than tenfold lower than for the glycolytic enzymes, since most of the kinases have K_m values that are below 100 µM. K_m values above millimolar would be more consistent for an emphasis on the speed of the reaction.

An important contrast emerges for the two groups in Fig. 2.3. Glycolytic enzymes generally have higher K_ms. Glycolysis is the highest flux pathway. Therefore, we see that K_m values are higher than average for these enzymes, to enable the turnover rates required. With these higher K_ms, the enzymes may frequently bind an incorrect metabolite, but they will not bind it tightly and will therefore release it almost instantly. And should they react chemically with it, the new compound produced may still have a use, since the cell has a variety of six carbon sugars, and their three carbon derivatives.

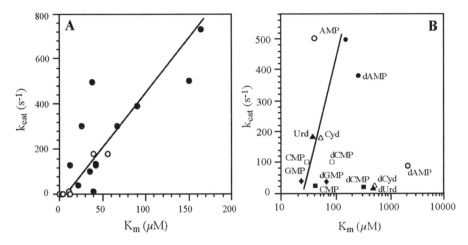

Fig. 2.4. Specificity of nucleoside and nucleotide kinases. (**A**) Nucleoside (*open circle*) and nucleotide (*filled circle*) kinases with values for their principal acceptor substrate. The best fit defines the specificity constant as 3.9×10^6 s^{-1} M^{-1}. (**B**) For some enzymes in (**B**), the data point for the principal substrate, a ribonucleoside or ribonucleotide is reproduced, along with the line denoting the specificity constant, as well as data for the deoxy analogs tested with these enzymes. The same symbol is used for data points for the same enzyme. Note that the scale for the abscissa has changed

With the kinases we see a spectrum of affinity values (Figs. 2.3 and 2.4). The highest K_ms for these are at 170 μM, which is below the average for the glycolytic enzymes, but many show much tighter binding. These latter are deoxynucleoside kinases, whose activity is not constantly needed, since most cells do not need a steady supply of deoxynucleotides at all times. Overall, we see a balance between the need for speed vs. the need for specificity and control. A guiding restraint is the need for an appropriate, or at least a minimal rate for each specific enzyme. Because tighter binding leads to a slower overall k_{cat} there is a limit to how discriminating an enzyme can be.

With these two opposing constraints, enzymes have evolved to have just the right affinity for their substrate. Due to the type of selection constraints described here, the concentration of the substrate is normally comparable to the affinity constants of enzymes that bind it. And the concentration of different substrates may vary over a range from micromolar to perhaps 10 mM, but again each substrate exists at a concentration appropriate for the rate at which it is consumed in however many pathways that require it.

2.3 Enzyme Specificity: k_{cat}/K_m

We can simplify (2.2) for the situation where the normal substrate concentration is low, since this would be the condition when an enzyme might more easily bind an available analog. At low [S] this term drops out of the denominator of (2.2) to give:

$$v = \frac{k_{cat}}{K_m}[E]_o[S].$$

(2.5)

The preceding discussion on defining the limits for rates and affinities of enzymes has established that these two features are related, and therefore an enzyme may achieve an ideal balance by optimizing the ratio of k_{cat}/K_m, which describes the enzyme's efficiency for any substrate or metabolite. This ratio is also known as the *specificity constant*. Since within a cell any enzyme may encounter a variety of molecules that are analogs of its normal substrate, either of these terms in the specificity constant may vary depending on how well the enzyme and metabolite interact. When both k_{cat} and K_m change correspondingly, the specificity constant is not varied. But, in vitro one might demonstrate, that some analog binds with a K_m 100-fold higher than for the normal substrate, but with the same k_{cat}. Then the specificity constant for the analog is 100-fold lower. The specificity constant then permits a meaningful quantitative comparison for an enzyme's ability to chemically react with a substrate.

2.3.1 A Constant k_{cat}/K_m may Permit Appropriate Changes for Enzymes with the Same Enzyme Mechanism

To see the effects of the constraints discussed above, it is helpful to examine a single group of enzymes that all have the function of phosphorylating either a nucleoside or a nucleotide. These enzymes appear to be descended from a common ancestor,[87] but have diversified so that most of them are fairly specific for a single substrate, or sometimes for two similar substrates. As shown in Fig. 2.4a, they have also diverged in the affinity for their principal substrate, and concomitantly in their maximum rates. The separate data points in this figure are based on values from the literature for the different enzymes. The best fit to these data defines a line with a slope equal to k_{cat}/K_m, which has a value of 3.9×10^6 s^{-1} M^{-1}. This specificity defines this set of enzymes as completely normal or average for these values. A few of the sample data points in Fig. 2.4A deviate noticeably from the average values. These may be true outliers for which some special explanation may yet be obtained, but they may also reflect some variation in how the data were obtained by different laboratories and with changing technologies.

Although they are descended from a common precursor, we also see that the individual enzymes have in fact altered their affinities and therefore their rates, while maintaining the same specificity constant. The most discriminating enzymes are deoxythymidine kinase and deoxycytidine kinase. They have a K_m at 1–2 µM, and consequently are also very slow with a k_{cat} of just above 1 s^{-1}. At the other end are UMP kinase from pig, and AMP kinase from chicken. These enzymes have very poor affinities, with high K_ms at 150–170 µM, but are therefore significantly faster with maximum rates at 500–700 s^{-1}.

In Fig. 2.4A we again see the specificity that many of these enzymes have for their normal ribonucleoside or ribonucleotide substrates. These enzymes will also phosphorylate the deoxy versions of the normal substrates, but normally at a lower k_{cat}, despite a very much greater K_m. Only two enzymes, GMP kinase and one of the CMP kinases, have almost as good a rate with the deoxy substrate. The difference in specificity is normally greater for enzymes that already have a low K_m for the normal substrate.

There is a definite trend between the measured K_ms of the enzymes for their acceptor substrates, and the measured concentration of these substrates in different cells: AMP and UMP normally exist at >100 µM, while deoxynucleosides are below 1 µM.[31] If an enzyme such as deoxycytidine kinase has a substrate normally present at 1 µM or below, then it must have an appropriately lower K_m in order to discriminate for this uncommon substrate. Enzymes such as UMP kinase can afford higher K_ms, since their normal substrate is sufficiently abundant at a concentration above 100 µM. While this has not generally been measured, it would be logical for the high K_m, high k_{cat} enzymes to be present at lower concentrations, as long as their actual rate of catalysis is adequate for the conditions of the cell in which they function. Then, even though UMP kinase may also bind some of the other pyrimidine substrates in the cell, such as deoxythymidine or deoxycytidine, it will bind them more poorly, and because UMP kinase is itself at a lower concentration, it will not contribute much to the normal synthesis of dCMP or dTMP. Therefore, the varied kinetic properties for the enzymes in Fig. 2.4 are consistent with the cell being able to have enough control for the formation of each nucleotide.

2.3.2 The Specificity Constant may Apply to Only One of the Two Substrates for a Group of Enzymes with the Same Mechanism

The kinases in Fig. 2.4 are named for the acceptor substrate, to which the phosphate group will be transferred. And as we see in Fig. 2.4, there exists the same specificity constant for all the acceptor substrates of this set of related kinase enzymes. Since all these enzymes use the same phosphate donor substrate, ATP, it is interesting to note that they have no constant specificity for ATP, as shown in Fig. 2.5. This figure shows no common feature for the use of ATP, though K_m values are mostly below 200 µM. While it is logical for these kinases to show discrimination for their preferred acceptor substrate,

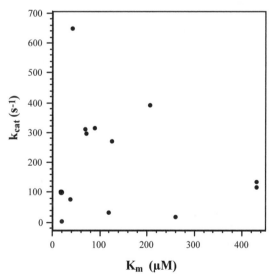

Fig. 2.5. Specificity of nucleoside and nucleotide kinases for ATP

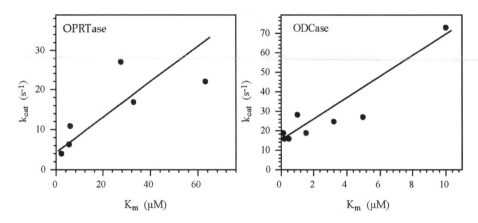

Fig. 2.6. Specificity constants for the same enzyme activity, for enzymes from different organisms. OPRTase, orotate phosphoribosyltransferase; ODCase, OMP decarboxylase

it is not necessary for them to show a comparable affinity for ATP, though this might be expected given that these enzymes are related to the same ancestor.

Again, it is also worth noting that the affinity for ATP is almost as strong as it is for the acceptor substrates. There presumably is no stringent need for these enzymes to show a preference for the phosphate donor. In terms of the chemical reaction for a kinase, any nucleoside triphosphate (NTP) would be an energetically equivalent donor substrate. And studies with uridine kinase have shown that this enzyme does not discriminate at the catalytic site between ribo-NTPs and deoxyribo-NTPs, and also accepts purine and pyrimidine NTPs.[74] Such results are consistent with the phosphate donor site of this enzyme being mostly occupied by the three phosphate groups, since little binding discrimination is evident for the ribose or the base.[88] One might then expect a very high, nondiscriminating K_m for ATP, and uridine kinase does have the highest K_m for ATP in Fig. 2.5. Most of the other enzymes show a better affinity, suggesting again that some degree of discrimination is normally needed even for the phosphate donor.

ATP is one of the most abundant metabolites in cells, normally having a total concentration of 2.5 mM or higher.[31] One might then expect that kinases could have quite a high K_m for ATP, since they would always be able to bind it well enough. However, an average cell has at least several thousand kinases, for a total concentration of these ATP-binding enzymes of perhaps 2 mM. Then, the actual free concentration of ATP is perhaps only 0.5 mM, or even lower. If most kinases should be more active only when the cell ATP pool is abundant, then their affinities for ATP should be consistent with such available ATP concentrations. This hypothesis is consistent with the otherwise surprising data that kinases generally have low K_ms for ATP.

However, if discrimination for a phosphate donor is not in fact necessary, then the variation that is observed may simply be a concomitant result as these enzymes have evolved their separate specificity for the primary acceptor substrate. That is, a mutation leading to a desired change in affinity at the acceptor site, may have a modest influence on the adjacent ATP binding site, leading to a variety of affinities for ATP that are still good enough for normal phosphotransfer reactions.

2.3.3 The Same Enzyme Can Maintain Constant Specificity While Adapting to Changes

We saw in Fig. 2.4 that a group of enzymes with the same type of reaction can have a constant specificity for their acceptor substrate, while still varying in their specific rates and affinities. The exact same flexibility is also evident if one examines a single specific enzyme reaction. Figure 2.6 shows such results separately for the enzymes orotate phosphoribosyltransferase (OPRTase),[36, 89–94] and OMP decarboxylase (ODCase).[36, 89, 94–99] Based on sequence alignments the OPRTases come from a common ancestor, as do the ODCases.[100]

For both of the enzymes in Fig. 2.6 there is a greater than tenfold range in the affinities for the principal substrate when enzymes from different organisms are compared. These variations in affinity may then reflect some differences in the need for how discriminating the enzyme needs to be in whatever cell it serves. For both enzymes in Fig. 2.6, those examples with the lowest values for k_{cat} and K_m are from mammals. If one interprets this sample set from microbes to humans as an evolutionary continuum, these results would support the interpretation that discrimination is more important than speed for these two enzymes. This is then an interesting evolutionary choice, since these two enzymes have activity rates at the low end of the range for such values.

2.3.4 The Limits to k_{cat}/K_m

The formulation of the specificity constant allows this value to be highest when either k_{cat} is maximized or when K_m is lowest. As a simple illustration of this, let us use the extreme limits of k_{cat} and K_m for calculating a specificity constant of 10^7 s^{-1} M^{-1}:

$$\frac{k_{cat}}{K_m} = 10^7 \, \text{s}^{-1} \, \text{M}^{-1} = \frac{10^7 \, \text{s}^{-1}}{1 \, \text{M}} = \frac{1 \, \text{s}^{-1}}{10^{-7} \, \text{M}}.$$

This is intended to illustrate the range for either of the two variables in this relation. For most enzymes, a balance between these two extreme positions is observed. It does emphasize the point that high specificity not only is provided by the obvious high affinity of a low K_m, but may also be produced by a very poor K_m when that leads to an exceptional k_{cat}. Examples of this diversity are shown in Table 2.4.

For natural enzymes, the efficiency is normally $\geq 10^5$ s^{-1} M^{-1}. But for artificial enzymes, such as DNAzymes and abzymes, the specificity constant is normally at 10^3 s^{-1} M^{-1} or much lower. The efficiency for the DNAzyme in Table 2.4 approaches the lower range for normal enzymes. Although it is still very slow, it is quite an achievement for the scientists who constructed it. The abzyme shown is also one of the most efficient artificial enzymes developed, but since it only has to increase the activity by 10^6 over k_{non}, this is not that difficult a chemical reaction.

Table 2.4. The range of observed specificity constants

Enzyme	Substrate	k_{cat} (s^{-1})	K_m (M)	k_{cat}/K_m (s^{-1} M^{-1})	Refs.
4-Oxalocrotonase tautomerase	2-Hydroxymuconate	2.9×10^6	1.9×10^{-4}	1.5×10^{10}	37
Superoxide dismutase	$O_2^{\cdot-}$	1×10^4	1.3×10^{-3}	8×10^8	56
Carbonic anhydrase	CO_2	1×10^6	1.2×10^{-2}	8×10^7	3
Catalase	H_2O_2	4×10^7	1.1	4×10^7	1
Uridine kinase	Uridine	180	4×10^{-5}	4×10^6	74
Orotate phosphoribosyltransferase	Orotate	4	$2 \times 10{-5}$	2×10^5	94
β-Alanine synthase	NCβA[a]	0.6	9×10^{-6}	7×10^4	101
Abzyme	Nitrobenzisoxazole	0.66	1.2×10^{-4}	5×10^3	102
DNAzyme	ODC RNA[b]	2×10^{-4}	6×10^{-7}	3×10^3	103

[a]N-carbamoyl-β-alanine
[b]Ornithine decarboxylase mRNA

Since the specificity constant, as a second order constant, cannot exceed the rate of diffusion that governs the encounter of two molecules, values $\geq 10^8$ s^{-1} M^{-1} are normally interpreted as indicating near perfection for such enzymes. In a general sense, we might assume that only a little mutational fine tuning is needed to adjust any enzyme to have somewhat better k_{cat} or K_m, and thus to approach this plateau of perfection. It is quite likely that for many enzymes this will remain an unattainable limit. A limiting feature that is frequently unappreciated is the actual difficulty of the chemistry for some reactions. Evidence for this is in Fig. 1.2, where we see that for some reactions, the uncatalyzed chemistry is incredibly slow, because it is so difficult. Considering the architecture of most catalytic sites, it is almost standard for two or three amino acid residues to participate in the actual chemistry, as opposed to the binding of the substrate. Frequently a metal cofactor or an organic cofactor may also be involved when they provide an appropriate benefit.

Most enzymes have three amino acids that participate in the reaction chemistry.[111] While two amino acids, or even one, might be enough for some types of chemistry, with

Table 2.5. Kinetic rates for ribozymes[a]

Enzymatic reaction	k_{cat} (s^{-1})	K_m	k_{cat}/K_m (s^{-1} M^{-1})	Refs.
Natural ribozymes				
RNA cleavage	0.2	20 nM	10^7	104
RNA cleavage	0.0017	1 nM	1.5×10^6	5
Self-splicing intron	0.001	1 nM	9×10^5	105
RNA cleavage	0.004	43 nM	9×10^4	106
Peptide bond formation	5	5 mM	10^3	107
Engineered ribozymes				
RNA self-ligation	1.1	9 μM	1.2×10^5	108
Aminoacyl esterase	0.1	450	2.2×10^5	109
RNA self-cleavage	0.023	49	4.8×10^5	110

[a]When more than one RNA was studied, only the most active is listed

three amino acids the active site will always assure that the substrate binding has the correct chirality. However, with three or more amino acids, a limited number of special arrangements are possible for these catalytic agents, and the perfect three-dimensional organization may not be available for all chemical reactions. And even when it is achievable, the process of natural selection appears generally to have been satisfied with enzymes that have not attained this ideal of perfection. In this sense, we may appreciate those enzymes with the highest specificity constants, without expecting this to be a standard that most enzymes will achieve.

2.3.5 Ribozymes and the RNA World?

We have increasing reports of RNAs that have a catalytic function.[112] Since such RNAS are found in so many different species, their existence is frequently used to support a model for an "RNA world,"[113, 114] to signify a time before proteins had appeared, and when RNAs were the principal molecules for both catalytic functions and information storage. This model proposes that proteins came later since they require ribosomes to be synthesized, and ribosomes contain many RNAs. DNA also appeared later, and being much more stable, it then assumed the storage of information. Since proteins are more complex and versatile they emerged to take on almost all catalytic functions. Current ribozymes are seen as the vestiges of an early more complex RNA world.

One serious difficulty with the RNA world hypothesis is that natural ribozymes have very limited catalytic functions, with the majority only able to cleave or ligate phosphodiester bonds. They are also very slow catalysts (Table 2.5), seldom having k_{cat} values greater than 0.1 s^{-1}. These features may be explained by assuming that the RNA world was slower, and that other catalytic functions for RNAs had also existed, but have disappeared as protein enzymes replaced them. But, since we have many examples of very efficient protein nucleases, why have the currently existing ribozymes also not been replaced by the more efficient protein enzymes?

An attractive answer is that currently existing ribozymes generally function as riboswitches to control transcription of a gene, or the processing of its transcript.[114] Since such functions are not directly involved in maintaining a steady state level of some metabolite, speed is not as critical as discrimination for the correct bond to cleave. When compared to protein regulatory switches (Fig. 2.1), these riboswitches have similar kinetic qualities, and there would be no benefit to a cell for replacing them with proteins that could not do the job any better. Base pairing provides the most direct means for binding to a specific site on a nucleic acid, and we see that the ribozymes/riboswitches mostly have binding constants in the low nanomolar range (Table 2.2), much tighter than some of the protein transcription factors. There is no surprise that many RNAs have been employed for such a function.

The literature promoting an RNA world is too extensive for a proper discussion here. The second serious difficulty with this hypothesis is: how was RNA produced without enzymes? While this question also applies to enzymes or proteins in general, it is worth noting that there is sound experimental evidence for the formation of amino acids and polypeptides in an abiotic system, with only ammonia, carbon monoxide, and metal catalysts (Fe, Ni, and Na$_2$S).[115, 116] This provides evidence for the spontaneous formation of polypeptides, requiring only the simplest of starting compounds and conditions, that are comparable to what should have been available in the earliest abiotic seas. We have no such demonstration for an abiotic synthesis of polynucleotides.

Many metals would have been available in the abiotic seas, and they would have influenced the emergence and diversification of protein enzymes in two phases. Those polypetides or small proteins that were able to fold and bind to an available metal would have been stabilized by such binding, making them more abundant. Therefore, simply due to the stabilizing benefit of binding a metal, metalloproteins would have become more widely established in this initial phase. Since metals contribute to the chemistry of so many reactions, some of these new metalloproteins would have had an enzymatic function, and this would then have led to the natural formation of a diverse mixture of simple metalloenzymes. This suggestion is supported by the fact that the majority of currently characterized enzymes are metalloenzymes.

Once such simple catalysts are present, more complex molecules including RNAs can then be produced in a steady manner. While RNA clearly preceded DNA in early life forms, the available research data suggest that the earliest biological world must have included an abundant mixture of simple enzyme catalysts. While some ribozymes should have existed at this time, they were not the unique species for catalysis. Initially RNA would have become important for storing genomic information, while proteins continued to evolve into better catalysts.

As life forms continued to evolve, better methods for the regulation of metabolism would have enhanced the survival of such cells. Such improvements in the control of metabolism would have largely involved proteins to produce ever more complex allosteric regulatory enzymes. However, the process of life is opportunistic. The appearance and continued use of riboswitches as transcription factors provides a natural and logical benefit to their cells.

The one clear exception among the ribozymes is the ribosome which has a rate for peptide bond formation of 5 s^{-1} (Table 2.5). Since the atomic structure of the large ribosomal subunit shows only RNAs at the catalytic center, then the ribosome is also a ribozyme.[117] A distinctive feature for the ribosome is that it has a very modest affinity for the peptide, which permits more rapid turnover, although the k_{cat} is still near the lower limit for a metabolic enzyme (Fig. 2.1). Note that the value of k_{cat}/K_m is $10^3 \text{ M}^{-1} \text{ s}^{-1}$, making the ribosome one of the least efficient enzymes.

The need for a certain level of accuracy in the process of translation precludes that this mechanism should go very rapidly. Since the other two components of the translational complex are mRNA and tRNA, then the possibility of specific alignments is a direct benefit in having rRNAs as the central catalytic reactants of the ribosome. As an example, the 23S rRNA can bind to the conserved CCA terminus of any tRNA.

It must also be noted that ribosomes contain more than 50 proteins, and these had to exist before the ribosomal translation process had become standard. It is therefore plausible that the ribosome emerged in an early "protein world" where the above benefits of using RNAs made the RNA–protein complex of current ribosomes a successful catalyst to mediate the translation of messenger RNAs.

3

ENZYME KINETICS

Summary

Kinetic equations and ligand binding equations may be very similar. The four most widely employed kinetic formats are the Michaelis–Menten, Lineweaver–Burk, Eadie–Hofstee, and Hill. The *effective* binding range for a ligand with constant affinity is nearly 100-fold. At $[L] \leq 0.1K_d$, binding becomes ineffective. At $[L] \geq 10K_d$, binding approaches saturation. This effective binding range is easily remembered as the *2-log rule*. The effect of positive cooperativity is to make the binding range narrower, so that saturation is approached below $[L] = 10K_d$, by changing the affinity for the ligand. The effect of negative cooperativity is to extend the binding range so that saturation is never approached under physiological concentrations of the ligand.

3.1 Time Frames for Measuring Enzyme Properties

The complete measurement and description of all time-dependent properties of enzymes requires instruments and techniques that have a variety of recording speeds. Figure 3.1 depicts broad ranges of time that have been measured for many different enzymes. Ligand binding normally occurs in nanoseconds. Protein folding for single domains, and conformational changes between folded domains of a protein, or subunits within an oligomer, are normally in the submillisecond to millisecond range. Reaction rates for enzymes are in the microsecond to second time range.

The appropriate speed and sensitivity that may be used is then a function of what feature is being measured. For any presteady state process to be evaluated, various stopped flow or quenched flow techniques are possible. In such methods separate syringes containing enzyme and substrates are simultaneously emptied into a combined chamber to initiate substrate binding and the enzymatic chemical reaction. As the reaction mixture flows along capillary tubing into a spectrophotometric recording device, the

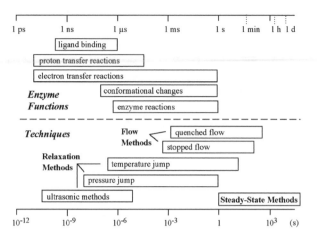

Fig. 3.1. The range of time scales for different properties of enzymes and for different methods

extent of the reaction is recorded, and it is directly related to the time for the solution to flow past. Relaxation methods enable one to rapidly perturb a preexisting equilibrium which then approaches a new state. In contrast, steady-state methods are on a slower time scale, and give specific information about overall enzyme rates (k_{cat} or V_{max}) and about the affinity for ligands that are bound (K_m or K_d).

3.2 Steady-State Kinetics

An enzymatically catalyzed chemical reaction may be measured at three different states. One may make measurements of the formation of product with time shortly after the components have been mixed together. Because the starting reagents are abundant, and no significant product has yet accumulated, the measurements are interpreted to reflect only the forward reaction. Since this system is still far from equilibrium, and if observations are made for a suitably limited time period, the system is defined as being at a *steady state*. This is the most common format for measuring enzymatic rates and affinity constants. While such measurements are almost always done in a test tube containing the immediate reagents plus the enzyme, this is understood as providing only an approximation for the same enzymatic reaction in a cell.

One may make very rapid measurements immediately after mixing, before the system has approached steady state. This is then a *presteady state* kinetic experiment. Such rapid measurements may be used to determine the rates for individual steps in the overall process, and it can help to define an enzyme mechanism. One may also allow the system to attain *equilibrium*, and by measuring the concentrations of both substrates and products, the equilibrium constant for the reaction may be calculated, and also the free energy for the overall process.

The cell is seen as a steady-state system, since cells are dynamic, with metabolites steadily being synthesized to replace the ones that are continuously consumed in other reactions. The important aspect of such steady-state systems is that there is a net flux of metabolite consumption or production in one direction and this is what the observer measures.

3.2.1 The Meaning of v and k_{cat}

Although most enzymes use two substrates in their normal reactions, it is standard to evaluate kinetic or binding constants between the enzyme and only one of these substrates in a single experiment. This is represented in either of the two versions of the standard scheme:

$$E + S \underset{k_{off}}{\overset{k_{on}}{\rightleftharpoons}} ES \overset{k_{cat}}{\longrightarrow} E + P$$
$$E + S \underset{k_{-1}}{\overset{k_1}{\rightleftharpoons}} ES \overset{k_2}{\longrightarrow} E + P$$

$$(3.1)$$

It has become standard in enzymology to refer to the speed or velocity of an enzyme reaction as v, and therefore the maximum velocity as V_{max}. For much of the twentieth century the units for v were μmol/min. This was practical because the activity observed was often measured by some spectrophotometric technique. Since a single cuvette normally required 2–3 ml of assay volume, the entire reaction was often done in volumes of 20–100 ml, and in this volume the reaction normally generated micromoles of product. By 1980 micropipets had become standard, and combined with the ready availability of isotopically labeled metabolites, an economy of scale evolved to reduce assay volumes to 100 μl or less. For these reduced assay volumes nanomoles of product are more routine.

Since the amount of product measured is a function of both the reaction time, and the concentration of enzyme, many scientists reported their results as the *specific activity*, which is the measured activity per mg of enzyme protein. This unit has more information, and enables other researchers to replicate any published measurements. The international *Unit* for enzyme activity is set to be 1 μmol/min/mg enzyme. This is equivalent to the often used nmol/min/μg enzyme. It is worth noting that for many years researchers had difficulties in purifying their enzymes, and frequently used preparations that were far from pure. Therefore, it was unclear what fraction of the protein in an enzyme assay actually represented the enzyme being measured.

When comparing specific activities for very different enzymes, this standard unit is not ideal. As a simple example, 1 mg of lysozyme (M_r 14.2 kDa) has almost seven times the number of enzyme molecules as 1 mg of glycogen phosphorylase (M_r 96.7 kDa). Toward the end of the twentieth century enzyme purification had progressed so that purity for enzyme preparations was now expected. Knowing that the enzyme was pure enabled the direct evaluation of the number of catalytic reactions per enzyme molecule. When this is done under optimal conditions it produces the value for V_{max}, and for a single enzyme molecule this is defined as k_{cat}, and has units of reciprocal time. This value is also referred to as the *turnover number*. We can directly compare k_{cat} values for any enzymes.

An important point regarding k_{cat} is that it represents the maximum rate for the enzyme, under optimal physiological conditions, and with a saturating concentration of substrate(s). The cellular concentration of any substrate seldom even approaches saturation, and therefore cellular rates for enzymes are more likely to be between 30% and 70% of V_{max}.

3.3 The Most Common Graphic Plots

3.3.1 The Michaelis–Menten Plot

The Michaelis–Menten equation was originally published in 1913,[45] as a direct derivation for the equilibrium between free enzyme, substrate, and the E–S complex. In this seminal paper these authors proposed several important procedures for the study of enzyme kinetics. They initiated the use of a buffer for enzyme storage and enzyme assays, and proposed that kinetic values should be obtained from the initial velocity measurements, before the substrate is significantly lowered, and before product inhibition can become significant. Since the authors were studying the kinetics of invertase on sucrose, there was only a single substrate to be considered, leading to their equation:

$$\varphi = \Phi \cdot \frac{[S]}{[S]+k}, \tag{3.2}$$

where φ is ES, Φ is E_{total}, and k is the affinity constant. Since the measured activity, v, is directly proportional to $[ES]$, and V_{max} is proportional to E_{total}, this equation is more normally presented as:

$$\frac{v}{V_{max}} = \frac{[S]}{[S]+K_m}. \tag{3.3}$$

By simple inspection of (3.3), it is evident that when the substrate concentration, $[S]$, is equal to the affinity for the substrate, K_m, then the observed rate must be equal to one half of the maximum rate.

When $[S] = K_m$, $v = 1/2V_{max}$.

The measured velocity curve for the Michaelis–Menten equation is shown in Fig. 3.1. The curve is calculated for a data set with 20 sample points, at regular intervals of substrate concentration. V_{max} has been defined at 0.1 nmol/min, and the K_m at 10 μM. Inspection of the Michaelis–Menten plot (Fig. 3.1) makes an immediate problem evident. Although the substrate concentration has been increased over a 100-fold range, V_{max} has not been achieved. In an actual experiment with some scatter in the data points, one would not be certain about the actual position of V_{max}, and this would limit the accuracy for determining $1/2V_{max}$, which is needed to define the K_m.

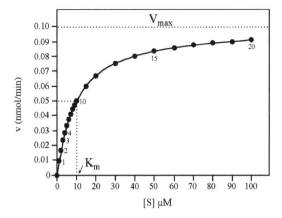

Fig. 3.2. The Michaelis–Menten plot as currently used, showing the kinetic rate curve for an enzyme catalyzed reaction when the enzyme has constant binding affinity for the substrate. The curve is calculated for the concentrations shown, at the given V_{max} of 0.1 nmol/min and K_m of 10 µM

It is important to note that Michaelis and Menten realized the difficulty of obtaining V_{max} and K_m from such a hyperbolic plot, and in fact did not show their data in a plot of the form shown in Fig. 3.2. They displayed their data in a semilogarithmic plot, as shown in Fig. 3.3, and described how to fit a straight line to the most linear portion of the curve, so that the value at the midpoint must then give the K_m. Despite the great success of the Michaelis–Menten equation for analyzing kinetic data, this semilogarithmic plot has not been widely employed.

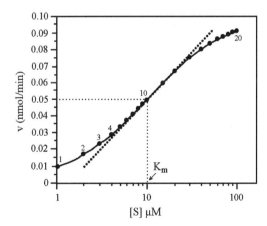

Fig. 3.3. The Michaelis–Menten plot in the format that was originally published, showing the kinetic rate curve for an enzyme catalyzed reaction when the enzyme has constant binding affinity for the substrate. The same data set from Fig. 3.2 is used

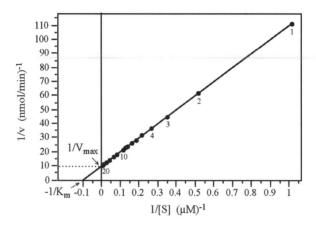

Fig. 3.4. The Lineweaver–Burk plot, showing the kinetic rate curve for an enzyme catalyzed reaction when the enzyme has constant binding affinity for the substrate. The same data set from Fig. 3.2 is used

3.3.2 The Lineweaver–Burk Plot

The semilog plot (Fig. 3.3) was not widely used, and scientists more routinely chose to display results in hyperbolic plots with the format of Fig. 3.2. This resulted in difficulties with the accuracy for the determination of V_{max} and K_m, especially when the experiment was not performed over an extensive range of $[S]$. A solution to this difficulty was developed in 1934 when Lineweaver and Burk transformed the Michaelis–Menten equation into the equation for a straight line.[118]

$$\frac{1}{v} = \frac{K_m}{V_{max} \cdot [S]} + \frac{1}{V_{max}}.$$
(3.4)

The kinetic plot defined by this equation is shown in Fig. 3.4. For the plots in this chapter, I have chosen to display results for 20 sample points, at regular substrate concentration intervals, and over a 100-fold range of $[S]$. However, in the normal literature authors frequently show experiments with five to eight data points, over a tenfold to 20-fold concentration range. For such results, a straight line may readily be extrapolated to the y-ordinate, to determine V_{max}. This value is given as equal to 0.1 nmol/min in Fig. 3.2, and the extrapolated value for the reciprocal ($1/V_{max}$) is equal to 10 in Fig. 3.4. Since the slope for the line defines V_{max}/K_m, the line may also be extrapolated to the abscissa to obtain the negative value for $1/K_m$. While the ability to generate a straight line is a benefit, this plot also has some disadvantages. The same set of 20 data points are shown in Figs. 3.2 and 3.4, and it can be seen that the Lineweaver–Burk plot compresses the data points along the x-axis. Whether the straight line is drawn to the points by eye, or by a programmed curve fit, points 1 and 2 receive unusual weight. Since these are the data points at the lowest $[S]$ values, they tend to have higher error values than the points at higher $[S]$, and therefore introduce an unwanted error in how the fit is made for the line, which can bias the interpreted values for K_m and V_{max}.

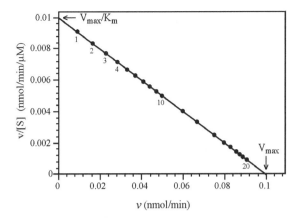

Fig. 3.5. Eadie–Hofstee plot, showing the kinetic rate curve for an enzyme catalyzed reaction when the enzyme has constant binding affinity for the substrate. The same data set from Fig. 3.2 is used

3.3.3 The Eadie–Hofstee Plot

To correct for any potential bias introduced by the variable error in the measured rate values, a plot was devised that graphed $v/[S]$ vs. v, one that would be statistically more robust and give the best result for K_m and V_{max} (Fig. 3.5). At least four different scientists derived an equation for this type of plot over 20 years, always publishing in major journals. As Table 3.1 shows, the actual equations in the original articles do not immediately appear to be the same, and this graphic format did not acquire an instant popularity. Therefore succeeding researchers appeared to be unaware of the earlier results, and again derived this format independently. This has resulted in this graphic plot being named for one or more of these contributors as the Eadie–Hofstee plot,[119, 120] or the Scatchard plot,[121] etc. While C.S. Hanes was actually the first to derive this equation, he did not present data in a plot of $v/[S]$ vs. v. He simply derived the equation and showed tables with calculated data that he matched to the standard Michaelis–Menten plot.[122] I will use the name Eadie–Hofstee for this plot, since that is more widely used in the current literature.

Also, note that Scatchard was concerned directly with ions and small molecules binding to serum proteins, so that his published equation is a true binding equation. To be more consistent with the format for the current version of this plot of $v/[S]$ vs. v, the different equations of Table 3.1 are more frequently rearranged, to produce (3.5), which is then attributed in different textbooks or publications to one or more of the authors in Table 3.1.

$$\frac{v}{[S]} = -\frac{1 \cdot V}{K_m} + \frac{V_{max}}{K_m}.$$ (3.5)

The format of (3.5) is to make it apparent that the actual graphic plot will be $v/[S]$ vs. v. For simplicity this equation can also be written as:

Table 3.1. Original equations for the Eadie–Hofstee Plot

Author (year)	Published equation	Interpreted version	Refs.
Hanes (1932)	$\dfrac{a}{v} = \dfrac{a + K_S}{V_{\infty}}$	$\dfrac{[S]}{v} = \dfrac{[S] + K_m}{V_{max}}$	122
Eadie (1942)	$V = V - K_p c$	$v = V_{max} - K_m\left(\dfrac{v}{[S]}\right)$	119
Scatchard (1949)	$\bar{v}/c = k(n - \bar{v})$	$\dfrac{B}{[S]} = \dfrac{1}{K_m}(B_{max} - B)^a$	121
Hofstee (1952)	$V_m = v + (v/S){\cdot}K_m$	$V_{max} = v + \left(\dfrac{v}{[S]}\right)K_m$	120

$^a B$, fraction of enzyme binding ligands, comparable to v

$$\frac{v}{[S]} = \frac{V_{max} - v}{K_m}. \tag{3.6}$$

Equation (3.6) may be derived from each of the equations in Table 3.1, and the format for this plot is shown in Fig. 3.5.

The resulting plot (Fig. 3.5) is designed to be linear, and easily leads to either K_m or V_{max}. Of the three graphic formats (Figs. 3.3–3.5), the Michaelis–Menten plot continues to be the most widely used, even though it is the weakest for providing either K_m or V_{max} with accuracy. The Lineweaver–Burk plot is fairly popular, but is more commonly used to display inhibition kinetics, since it gives visually distinct graphic patterns for the different types of inhibition that are possible. Despite its statistical superiority, the Eadie–Hofstee plot is used infrequently. These usage patterns may reflect the ease with which humans interpret a simple linear figure such as the Michaelis–Menten plot, and this also applies to the Lineweaver–Burk plot, even though the axes now display the reciprocal for v and $[S]$. Since the Eadie–Hofstee plot has no direct axis for $[S]$ alone, it may be more difficult to interpret for scientific readers who do not routinely use these graphic formats.

3.3.4 The Hill Plot

An additional plot had been formulated much earlier by A.V. Hill in 1910 to describe the binding of oxygen to hemoglobin.[123] Researchers at that time were unsure as to the oligomeric state of hemoglobin, and Hill therefore devised an equation with a constant, n, to represent any possible degree of association of hemoglobin molecules from monomer to tetramer or higher:

$$y = 100\frac{Kx^n}{1 + Kx^n}. \tag{3.7}$$

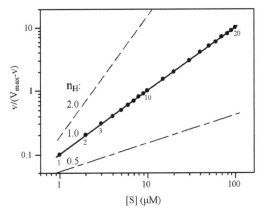

Fig. 3.6. A Hill plot, showing the kinetic rate for an enzyme catalyzed reaction when the enzyme has constant binding affinity for the substrate. The same data set from Fig. 3.2 is used. The sample data set has a slope of 1.0, and shows no cooperativity. The *dashed lines* are examples of positive cooperativity (a slope of 2.0) or negative cooperativity (a slope of 0.5)

In (3.7), y is the saturation of hemoglobin with oxygen, and equals $(HbO_2)/Hb_{total}$, which is then comparable to v/V_{max}. This relationship is evident in the fact that v represents the concentration of active enzyme molecules, and these must bind a substrate to initiate catalysis. And V_{max} is attained when all enzyme molecules are binding substrate. Also in (3.7) X is the $[O_2]$, comparable to $[S]$, and K is the binding *association* constant, which requires conversion to a dissociation constant, to be consistent with current usage. This original equation was then rearranged so as to quantify the extent of cooperativity in a kinetic experiment. The version of this equation that is more commonly used for this plot is:

$$\log \frac{v}{(V_{max} - v)} = n_H \log[S] - \log K_m. \tag{3.8}$$

The Hill coefficient, n_H, is the slope of the plot in Fig. 3.6, and it defines the extent of cooperativity. When the affinity of the enzyme for its substrate is constant, n_H will have a value of 1.0. If the affinity changes to become greater, this means positive cooperativity, and n_H will have values of about 1.5 or greater. As will be discussed in Chap. 5, the upper limit for n_H is the number of subunits in an allosteric oligomer. This upper limit for n_H is seldom observed. A value below 1.0 represents negative cooperativity, and indicates that affinity is becoming weaker. For negative cooperativity, n_H is almost never lower than 0.5.

3.4 Interpreting Binding Constants

In discussions of enzymes interacting with ligands one can do direct binding experiments, to obtain a true K_d, or different types of kinetic experiments to obtain a K_m or K_i. Let us first consider the general binding of any ligand, L, to an enzyme, E.

$$E + L \underset{k_{-1}}{\overset{k_1}{\rightleftharpoons}} EL. \tag{3.9}$$

As shown, this is an equilibrium expression for the association of L with E, and the equilibrium for this expression, (as written) would express an *association* constant (K_a). Biochemists almost always use the reciprocal expression to define a *dissociation* constant (K_d), which can be defined as:

$$K_d = \frac{k_{-1}}{k_1} = \frac{[E][L]}{[EL]}. \tag{3.10}$$

A major advantage of the latter is that it has simpler units, concentration, which one can determine directly from the expression shown in (3.10).

The Michaelis–Menten expression describes:

$$E + S \underset{k_{-1}}{\overset{k_1}{\rightleftharpoons}} ES \overset{k_2}{\longrightarrow} E + P \tag{3.11}$$

For which

$$K_m = \frac{k_{-1} + k_2}{k_1} \tag{3.12}$$

And

$$\frac{v}{V_{max}} = \frac{[S]}{K_m + [S]}. \tag{3.13}$$

The Michaelis–Menten expression is related to the direct binding expression, as shown by rearranging (3.10):

$$K_d = \frac{[E][L]}{[EL]} = \frac{[E_o - EL][L]}{[EL]} = \frac{[E_o][L]}{[EL]} - \frac{[EL][L]}{[EL]}. \tag{3.14}$$

E_o is the total concentration of enzyme, while EL is the concentration of enzyme bound with ligand. The difference between these equals the actual concentration of free enzyme, E. Continuing with (3.14), we can rearrange to obtain:

$$K_d + [L] = \frac{[E_o][L]}{[EL]}. \tag{3.15}$$

then,

$$\frac{K_d + [L]}{[L]} = \frac{[E_o]}{[EL]}. \tag{3.16}$$

This can be rearranged to provide:

$$\frac{[EL]}{[E_o]} = \frac{[L]}{K_d + [L]}.$$ (3.17)

Since $[EL]$ represents the concentration of enzyme involved in binding the ligand, and E_0 is the total concentration of enzyme, $[EL]/[E_0]$ is the fraction of the enzyme population that is actively binding ligand under whatever conditions, and this is then comparable to the definition for enzyme activity stated by the Michaelis–Menten equation:

$$\frac{v}{V_{max}} = \frac{[S]}{K_m + [S]}.$$ (3.18)

The formula for the equilibrium dissociation constant of any ligand, L, is quite similar to the expression for enzyme velocity, v, as a function of $[S]$ to yield the Michaelis constant, K_m, which is evident when comparing (3.17) and (3.18). Note that similar terms are used to express a dissociation constant: K_d or K_S. These represent true equilibrium constants. K_m is not a true equilibrium constant, but an initial "steady-state" constant, due to the catalytic step, k_2, which may also be defined as k_{cat}. Thus, for this system, the "pseudoequilibrium" constant, K_m, is given by:

$$K_m = \frac{k_{-1} + k_2}{k_1} = \frac{k_{-1}}{k_1} + \frac{k_2}{k_1} = K_d + \frac{k_2}{k_1}.$$ (3.19)

Since catalysis is frequently rate limiting, k_2 is normally much slower than the association rate, k_1, and therefore $K_m \approx K_d$. Since the majority of enzymes are in this category, it is customary to treat the K_m as an equilibrium constant, though one must be aware that this is not correct for special cases.

3.5 Energetics of Enzyme Reactions

Up to this point we have discussed all kinetics in terms of the standard Michaelis–Menten model, with the assumption that the K_d for substrate binding is the same as the K_m for the overall reaction, and that k_{cat} is the same as k_2, as depicted in (3.1). The actual mechanisms and observed kinetic values for the majority of enzymes support these assumptions. However, there are two groups of enzymes that require an alternate interpretation, since they involve mechanisms that either may have a very low transition state barrier, or may have multiple ES intermediates. Models for all three groups will be examined below, so that the distinguishing features for each one will become more evident.

Michaelis–Menten model:

$$E + S \underset{k_{-1}}{\overset{k_1}{\rightleftharpoons}} ES \overset{k_2}{\rightarrow} E + P$$

$$k_2 \ll k_{-1}$$

$$k_{cat} = k_2$$

$$K_m = K_d$$

$$K_m = \frac{k_{-1} + k_2}{k_1} \approx \frac{k_{-1}}{k_1} = K_d$$

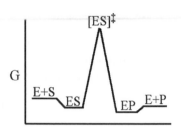

Fig. 3.7. Energy profile for a Michaelis–Menten scheme

3.5.1 Michaelis–Menten Model

The scheme associated with Fig. 3.7 shows the simplest scheme, where free enzyme and substrate bind to form an ES complex. The on rate, or binding rate, is represented by k_1. If binding is weak, this complex will readily dissociate to the free enzyme and unmodified substrate, for which the rate is k_{-1}. Some fraction of the ES complex will undergo chemical catalysis, for which the rate is k_2. The fraction calculated from the ratio of k_2 to k_{-1} is some number less than 1, and defines the *commitment to catalysis*. Therefore, after the substrate is bound, it is more likely for it simply to be released, without undergoing any chemical reaction, especially if the K_m is poor. But even enzymes with K_ms in the micromolar range will normally release the substrate more often than they will chemically convert it to the product. This is a simple consequence of a high K_m, since the poor affinity of the ES interaction permits k_{-1} to be faster than k_{cat}.

Key points for this model are that there exists a significant energy barrier for catalysis, and therefore k_2 will be moderate or even low. Since there are no significant intermediates other than the ES complex itself k_{cat} reflects the limiting catalytic step, and equals k_2. Most currently characterized enzymes are consistent with this model.

3.5.2 Briggs–Haldane Model

There are chemical reactions where there is no great energy difference between reactants and products, and no significant energy barrier for the catalytic reaction.[124] This means that the transition state is less than the energy required for E and S to collide in solution. The off rate for the bound substrate may be very fast. Since there are no other significant intermediate species, k_{cat} may equal k_2. Then k_{cat} will be very fast, and may approach or equal k_{-1}, the off rate for binding of substrate with enzyme. When calculating K_m for the Michaelis–Menten model k_{cat} is normally negligible, relative to k_{-1}, and does not contribute to the calculated value for K_m. In the Briggs–Haldane model, with k_{cat} approaching k_{-1}, K_m may approach $2K_d$. Enzymes with this mechanism include the fastest enzymes known, such as carbonic anhydrase, catalase, and triose phosphate isomerase (Fig. 3.8).

Briggs–Haldane model:

$$E + S \underset{k_{-1}}{\overset{k_1}{\rightleftharpoons}} ES \overset{k_2}{\rightarrow} E + P$$

$$k_2 \approx k_{-1}$$

$$k_{cat} = k_2$$

$$K_m \geq K_d$$

$$K_m = \frac{k_{-1} + k_2}{k_1} \approx \frac{2k_{-1}}{k_1} \approx 2K_d$$

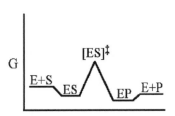

Fig. 3.8. Energy profile for a Briggs–Haldane scheme

3.5.3 Additional Intermediates Model

For certain reactions there are additional intermediates, which therefore alter the normal kinetic description. Examples of this category include enzymes whose mechanism involves formation of a covalent intermediate between the enzyme and the substrate. Alkaline phosphatase abstracts the phosphate group from various organic phosphate compounds, and transiently binds the phosphate as a covalent intermediate, which is then attacked by water to release inorganic phosphate. Although normally transient, the phosphoenzyme intermediate is sufficiently stable that it has been isolated, in support of such a mechanism for catalysis.[125]

Enzymes with such a covalent intermediate bind the first substrate, from which they abstract some moiety to be transferred, and release the remaining group as the first product (Fig. 3.9).

Additional Intermediates model:

$$E + S \underset{k_{-1}}{\overset{k_1}{\rightleftharpoons}} ES \underset{k_{-2}}{\overset{k_2}{\rightleftharpoons}} ES' \overset{k_3}{\rightarrow} E + P$$

k_3 is slow

$$k_{cat} = \frac{k_2 \cdot k_3}{k_2 + k_3}$$

$$K_m < K_d$$

$$K_m = \frac{K_d}{1 + k_{-2}/k_2}$$

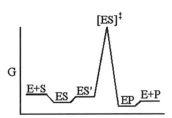

Fig. 3.9. Energy profile for a reaction with intermediates after ES

$$E + ATP \underset{k_{-1}}{\overset{k_1}{\rightleftharpoons}} E \sim P + ADP \rightleftharpoons NDP + E \sim P \underset{k_{-2}}{\overset{k_2}{\rightleftharpoons}} E + NTP \qquad (3.20)$$

They then bind the second substrate, the acceptor, onto which the group covalently bound to the enzyme will be transferred. Other enzymes in this category include many proteases that form an acyl-enzyme intermediate, and the acyl group is then transferred to an attacking water molecule. Many enzymes in this category are hydrolases. This permits a simplicity in the architectural design, since they have one binding pocket for the principal substrate, and must simply coordinate the ever available water molecule. Perhaps the simplest such enzyme is nucleoside diphosphate kinase (3.20), which is normally described as using ATP as a phosphate donor, and then transferring a phosphate to any other nucleoside diphosphate (NDP). This enzyme is actually quite promiscuous, and shows no binding preference for either NTPs or NDPs. The enzyme has one binding pocket for any NTP. The terminal phosphate is covalently bound to make a phospho-enzyme intermediate. The remaining NDP is then released. Into the same binding site any other NDP may now enter to receive the bound phosphate, and thereby become an NTP. Because of this back and forth shuttling of the moiety being transferred, this is referred to as a *ping-pong* reaction, or a double-displacement reaction.

4

PROPERTIES AND EVOLUTION OF ALLOSTERIC ENZYMES

Summary

The purpose of regulatory enzymes is to permit control of the rate at a specific metabolic step. There are two major groups of allosteric enzymes. Both employ a change in affinity for the substrate to alter activity. One of the groups experiences only a change in affinity for the principal substrate, while keeping the maximum rate fairly constant (*K-type* enzymes). The second group also demonstrates significant changes in affinity for the main substrate, and in addition has large changes in the maximum rate (*V-type* enzymes). The greatest changes in V_{max} are for enzymes that act as regulatory switches.

4.1 Different Processes for Controlling the Activity of an Enzymatic Reaction

Metabolic regulation is so important that every possible strategy has evolved and is actively used. These are summarized in Table 4.1, and will be briefly reviewed. Although I have used bacteria to establish the basic parameters for concentrations and kinetic rates, it is better to focus on a mammalian system to easily summarize the complete diversity of regulatory mechanisms. With bacteria mammalian cells share many forms of internal metabolite regulation. In addition, mammalian cells, as part of a larger organisms, have additional control modes by hormonal release.

An immediate and important distinction in the grouping of control processes is the time frame for action: those that will immediately *alter the activity of any existing enzyme* that is specifically targeted, and those that *alter the actual concentration of any specific enzyme*. These two main categories for regulation do not duplicate each other. They have distinctly different time frames for producing an effect, they normally have very different signals or stimuli to initiate the process, and they are produced by totally different mechanisms.

Table 4.1. Different processes for regulating an enzymatic step

A. Modifying the activity of an existing enzyme (occurs almost instantly)

    I. *Dynamic ligand binding equilibria*
       (1) Simple competitive inhibition: *at the catalytic site*
           • Produced by increased product, or product analog
           • Caused by changes in diet, metabolic state (stress, pregnancy, etc.)

       (2) Allosteric regulation: *at a regulatory site*
           • Produced by regulatory effectors (activators, inhibitors)

    II. *Covalent modification of the enzyme:* phosphorylation, adenylylation, etc.
           • Initiated by hormones or signal molecules

B. Modifying the quantity of enzyme

    I. *At the level of the gene* (occurs in hours to days)
       (1) Induction = $\uparrow$gene transcription (or $\uparrow$translation);
           • Initiated by hormones

       (2) Repression = $\downarrow$transcription
           • Initiated by hormones

    II. *At the level of the protein* (occurs quickly)
       (1) Proteolytic activation of inactive zymogen (proenzyme)
           • Initiated by various, specific stimuli

       (2) Catabolism or proteolytic degradation
           • An ongoing, constant feature
           • May also be increased by hormones

4.1.1 Modifying the Activity of an Existing Enzyme

The fastest response possible for any regulatory control is to influence those enzymes currently performing the targeted metabolic step. This can be accomplished by two different procedures, *ligand binding* and *covalent modification*. An example of the first regulatory process in Table 4.1, binding of the product as an inhibitor at the catalytic site, is an almost universal feature for enzymes. Since enzymes must have a catalytic site at which the product is made, then if this product increases in abundance it will naturally bind again at this site, leading to inhibition. This simple form of inhibition would have been the earliest, original mode that began to influence the rate of each and every enzyme.

4.1.2 Modifying Activity by Ligand Binding

Many enzymes have reversible reactions, so that, depending on the momentary concentrations of the reactants and products relative to the equilibrium position, the reaction may at times go in the reverse direction. But even if the reverse direction is energetically unfavorable, the actual presence of a product at the catalytic site will function as an inhibitor, and thereby diminish the physiologically forward reaction leading to more of this product. Any metabolite that resembles the product and is also able to bind at the catalytic site would also result in inhibition. And, if normal analogs exist that are sufficiently abundant, then this is again a constraint that would favor evolution of the catalytic site to be more discriminating against such inhibitors, and in favor of the normal substrate, consistent with the earlier discussion on the normal values for K_m and k_{cat}.

A subset of enzymes is allosterically regulated, and this signifies at least one regulatory site where an appropriate effector may bind. The immediate product is not routinely an allosteric effector, though we will see an example of this with hexokinase in Chap. 10. Almost all allosteric enzymes respond to at least one specific inhibitor, and this in most cases is the end product of the pathway controlled by this regulatory enzyme. Frequently there is also at least one activator. The activator may be the end product of a complementary pathway, and represents the availability of these additional components. Frequently ATP may be an activator, as a signal that the cell/organism is well nourished, and that therefore appropriate biosynthetic pathways may expand their activity.

A significant feature of this type of regulation is that it is completely sensitive to the concentration of the inhibitor or allosteric ligand. As the concentration of such a ligand increases relative to its K_d for the enzyme, this will cause a greater portion of the enzyme population to bind the ligand and display the change in activity produced by such binding. Similar to the earlier discussions for the range of K_m values, it is most logical that enzymes have evolved a sensitivity for regulatory ligands that is appropriate for the likely concentration of such ligands. Inside a cell, the activity of any enzyme will then be the natural result of its ability to bind substrates, which depends on [S] and K_m values, and of its ability to bind effectors, which depends on [L] and K_d values. Under normal conditions, this type of ligand binding by metabolites provides a major influence for maintaining the steady state of metabolism.

4.1.3 Modifying Activity by Covalent Modification

Since the extent to which an individual enzyme population is influenced to alter its activity is proportional to the concentration of the effector ligand, this type of control will always be somewhat limited. An alternate mechanism for regulatory control evolved that succeeds in keeping the affected enzyme indefinitely in a new conformation by covalently modifying it. The most widely used form of covalent modification is phosphorylation, and there are estimated to be greater than 1,100 protein kinases in the human genome (see Chap. 13). In addition to simply transferring the phosphate group from ATP to an enzyme to form a phosphoenzyme, we also have examples such as glutamine synthetase where the entire AMP moiety is attached to produce an adenylyl-enzyme. In a similar fashion, a regulatory enzyme for glutamine synthetase is itself controlled by uridylylation, the attachment of a UMP from the donor UTP.

The conformation produced by this type of alteration may lead to the enzyme being less active, or more active, as illustrated in Fig. 4.1. Ten allosteric enzymes are shown for this very limited region of central metabolism to be controlled by this mechanism. Since this regulatory mechanism is found in bacteria and all eukaryotes, it evolved quite early and has been adapted very widely.

Since the phosphorylated enzyme is stabilized in a conformation different from the native enzyme, it is necessary to have an opposing enzyme that reverses this action by

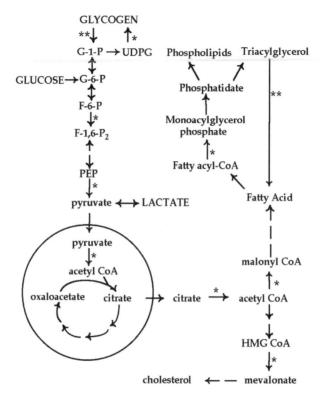

Fig. 4.1. Examples of enzymes regulated by phosphorylation. (*) enzymes inhibited by phosphorylation; (**) enzymes activated by phosphorylation

cleaving the phosphate group and restoring the enzyme to its native conformation. These are the protein phosphatases, and they also come in a large assortment, some of which have many phosphoenzymes as their potential substrate, and some of which are fairly specific for a single phosphoenzyme. To help illustrate the balance between these opposing kinases and phosphatases, Fig. 4.2 depicts the enzymes necessary for glycogen synthesis and glycogen catabolism.

The overall process illustrated in Fig. 4.2 shows a liver cell that has membrane receptors for binding the hormones glucagon and epinephrine. These hormones bind at very different receptors, but for simplicity only one receptor is shown. Binding of the hormone activates the membrane bound *adenyl cyclase* to form cAMP, which binds to the regulatory subunit of *protein kinase*, as shown at the top of Fig. 4.2. When the regulatory subunit (R) binds cAMP, it is stabilized in a new conformation that does not interact well with the protein kinase catalytic subunit (C). The independent protein kinase is now fully active, and able to initiate the phosphorylation cascade diagrammed (Fig. 4.2) by phosphorylating *phosphorylase kinase*, the protein kinase that specifically modifies *glycogen phosphorylase*, which is the enzyme that converts glycogen to glucose-6-P. This

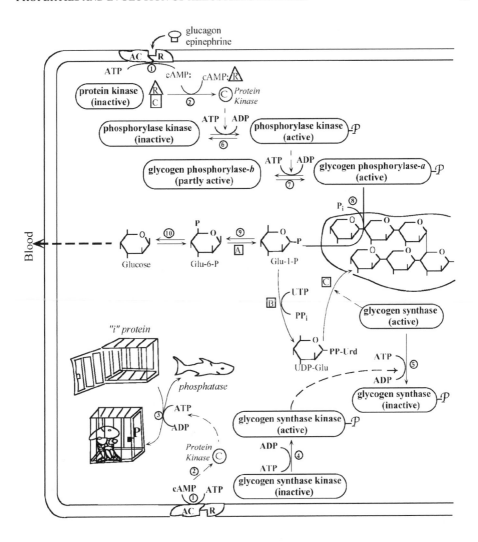

Fig. 4.2. The importance of enzyme phosphorylation in the control of glycogen metabolism: **(a)** in the membrane, AC = adenyl cyclase, R = receptor; **(b)** in the cytoplasm, C = catalytic subunit, R = regulatory subunit. "*i*" *protein* is a regulatory subunit that binds the *phosphatase*, comparable to R binding C of *protein kinase*. Note that when the "*i*" *protein* is phosphorylated, it binds and inactivates the *phosphatase*. Enzymes represented by *circled numbers* are involved in converting stored glycogen to glucose. Enzymes represented by *boxed letters* are in the opposing pathway for glycogen synthesis. Allosteric changes are produced by (1) binding of allosteric ligands or (2) covalent modification = E–*P* enzymes

same hormone initiated sequence at the bottom of this figure shows the protein kinase also phosphorylating *glycogen synthase kinase*, which in turn will phosphorylate, and thereby inactivate, *glycogen synthase*.

We see that enzymes can be activated by phosphorylation, or inactivated. Since the same signal initiates phosphorylation of enzymes that either make glycogen, or break it down, it is necessary that this covalent change causes one enzyme to be more active, and the other to be less active. Otherwise, a futile cycle of continued glycogen formation and breakdown would simply consume ATP with no other benefit.

It is also worth noting that while many of the enzymes in Fig. 4.2 are allosterically controlled by phosphorylation, two enzymes have separate regulatory subunits, and are inactive when their catalytic subunit binds to the regulatory subunit. The enzymes controlled in this fashion are the protein kinase, just described, and a general *protein phosphatase*. To emphasize the inhibitory effect of the regulatory subunit, near the bottom of the figure the *i protein* (for "inhibitor"), which controls the phosphatase is illustrated as a cage, and the active phosphatase as a shark, constantly seeking phosphoenzymes. Phosphorylation by protein kinase makes the *i protein* undergo a conformational change that enables it to bind and inactivate the phosphatase. With the phosphatase enzymes now inactivated, the phosphorylation cascade may continue and achieve the appropriate change in the different enzymes resulting in the net breakdown of glycogen to produce glucose that may be released into the blood.

Although most of the important enzymes in this metabolic scheme are controlled by covalent phosphorylation, the two initial opposing enzymes, protein kinase and protein phosphatase, are controlled by mass action, via binding of a regulatory subunit. This assures that the protein kinase remains active only so long as cAMP is abundant, since the cAMP will be cleaved steadily by a phosphodiesterase (not shown in Fig. 4.2). In a similar fashion, even though the *i protein* is activated by phosphorylation, it has a modest binding constant for the phosphatase, guaranteeing that all the phosphatase enzymes can never be inactivated.

If every phosphatase were bound by its regulatory subunit, then the enzymes that have been phosphorylated would remain in that state, resulting in a permanent change in their activity. This is clearly not desirable. The dissociation constant for binding of the phospho-*i protein* to the phosphatase is such that some phosphatase molecules will always remain free, and therefore active. Note that the phosphatase will convert the phospho-*i protein* back to the native *i protein*, thereby also increasing the number of active phosphatase molecules. They will then steadily cleave all the phosphates from the different enzymes that have been covalently modified, to restore the system to the original state.

A final point evident in Fig. 4.2 is that allosteric control by covalent modification and by ligand binding are not simply alternate mechanisms for a similar result. Covalent control, especially by an enzyme cascade as shown in the figure, is fairly fast and does not require the accumulation of multiple ligands. This is a very important feature, and is a clearly desirable mechanism inside a cell where osmotic pressure limits the concentration of effector molecules. However, the ultimate control is still by mass action, and depends mainly on the transient concentration of cAMP after the hormones have activated the membrane receptor and adenyl cyclase. Therefore, only a single effector molecule needs to change in concentration to produce a result involving six separate enzymes.

4.1.4 Modifying Activity by Altered Gene Transcription

The allosteric changes shown in Fig. 4.2 are designed to regulate those enzymes involved in the synthesis and breakdown of glycogen that are already in the cell. Normally this regulatory cascade would function for a brief time, until the appropriate blood glucose concentration is again attained. If a major change in the physiological state occurs, long term changes may be needed, and this more commonly involves a change in the total concentration of those enzymes that are important to that pathway. Examples of such changes in the state of an organism occur during growth, pregnancy, starvation or limited diet, serious illness, etc. To maintain a proper metabolic balance while responding to positive or negative stimuli or stressors for an extended time, the absolute concentration of many enzymes can be altered. Almost all stress situations result in changing hormone output, and some hormones have their effect by altering the transcription of genes for groups of enzymes or receptors. Both gene induction and repression occur. This balances overall regulation since the same hormonal stimuli may need to increase expression of one set of genes, while attenuating the expression of genes whose enzyme products would have an effect opposing the set that is being increased. For mammals such detailed regulatory control of gene expression makes it possible to increase or decrease the concentration of specific enzymes.

4.1.5 Modifying Activity by Proteolysis

Cellular proteins are constantly subject to damage by oxidation, radiation, hydrolysis, etc. When an enzyme is damaged and thereby nonfunctional, it needs to be efficiently removed, and replaced by a new copy. Therefore, damaged proteins are recognized by well-established mechanisms, so that they can be efficiently removed. The amino acids resulting from such proteolysis are then available for new protein synthesis, or during periods of fasting may be converted by deamination to glucogenic metabolites.

Proteolysis can also be the mechanism for activating an enzyme. This normally occurs with many proteases that are intended to function in a specific compartment, such as the digestive proteases in the stomach and small intestine, and these include pepsin, trypsin, and chymotrypsin. Other proteases, such as those involved in blood clotting, are intended to work only in response to a specific signal, but must constantly be available in an inactive state. This is a very logical strategy for controlling such potentially dangerous enzymes. By having the initially translated protein in an inactive form, such enzymes may be produced, with no damage to surrounding proteins in the cytoplasmic compartment where they are first synthesized.

For all these proteases the original protein made on the ribosome is a precursor that folds into an inactive conformation. Such enzyme precursors are called *zymogens*, since they will be converted to an enzyme. They are also known as proenzymes. The precursor conformation is inactive because an extended peptide segment of the enzyme normally blocks the active site, or prevents correct folding to produce an active site. When this peptide is removed by a specific protease, the resulting truncated protein then can form the proper conformation to have an accessible active site, so as to be functional at the time, or in the location, where its activity is required.

4.2 Evolving Allosteric Enzymes

The ability of allosteric enzymes to have altered conformations which lead to changes in binding or activity clearly distinguishes them from normal enzymes that lack these features. The most common viewpoint is that normal enzymes have but one conformation, which is the active form, and therefore are ever ready to perform their catalytic function. These enzymes maintain an activity level in response to the concentration of their own substrate that results by some level of catalytic flux in a metabolic pathway controlled by the regulatory enzymes. These normal enzymes always have enough activity to keep up with the pace set by the regulatory enzyme.

The regulatory enzymes, of course, are special. They have two or even more conformations, and they almost always have additional binding sites at which additional physiological effector molecules may bind to stabilize a conformation that is more active, or less active. And this provides the means for altered metabolic activity.

Our current knowledge of protein structures gives us a more realistic understanding of how this occurs. Based on studies of populations of a single type of protein molecule, it is now well established that any population of proteins will never all have a single conformation. There is in fact normally an extended range of folded states that any protein population will sample. States with the lowest energy will be heavily frequented, and states with higher energy levels will be less abundant. In reality therefore, any protein is actually able to have multiple conformations. However, along this range of potential folded states, they will not all be equally sampled, and therefore some states will be more abundant, some at modest concentrations, and some at very low concentrations.

For example, enzymes that are not allosteric, and that by kinetic measurements have a constant affinity and activity do not appear to give evidence of having any but that one active conformation. But, when such proteins are studied by NMR or by calorimetry, it becomes clear that they actually do have a range of conformations.[11, 12] This perception that any enzyme is an ensemble of conformations was clearly defined in a significant study by Sunney Xie and colleagues, in which they measured the activity of single molecules of β-galactosidase.[126] By making an artificial substrate that contained a portion that would become fluorescent upon being hydrolyzed by the enzyme, they had a convenient assay to measure the appearance of each product molecule. By attaching the enzyme to a bead, which in turn was attached to a glass slide, they were able to obtain beads with a single enzyme molecule, and then follow such an enzyme's catalytic rates for up to several hours. Although this enzyme has a k_{cat} of 730 s^{-1}, they could follow the activity on a submillisecond time scale, and observe frequent pauses when the enzyme was not active. The duration for such pauses lasted from milliseconds to seconds, values that are consistent with the time for movements within the enzyme tetramer of loops, or domains. And, because such individual enzyme molecules stochastically explored active and inactive conformations, the kinetics for a single enzyme molecule yielded the same values for k_{cat} and K_m as were obtained for a normal population of these enzyme molecules. This study then demonstrates that each enzyme molecule, as a function of time, explores all available conformational states, and an ensemble contains a distribution representing the frequency of all such states.

For nonallosteric enzymes, a majority of these molecules may be at or very close to the active state for any enzyme. Since the remaining protein states for this enzyme are not stabilized by any special effectors, then the most populated state, which is the active state, defines the activity for the enzyme population. In terms of kinetics it behaves as a single population, and the additional states that the enzyme may intermittently populate are not relevant to kinetic measurements. We may safely represent this enzyme as having but one conformation, even though we also have measurements showing the enzyme to briefly sample many conformations. The importance of this simplification about normal enzymes is that their active conformation is normally the most abundant form, although that is not essential. The key requirement for assuming a single conformation is that any additional conformations that have lowered activity are not variable within the conditions of the enzyme assay. If only 10% of the enzyme molecules are in an active conformation, with 90% inactive, this is functionally a single conformation if the ratio of the two conformation states cannot be altered. The total enzyme activity will always be due to those 10% that are in the correct conformation. Two or more conformations become significant only if some normal parameter of the assay influences the distribution among these conformations, since that will be observed as a change in K_m or V_{max}.

4.2.1 Allosteric Regulation by Stabilizing the Appropriate Species in an Ensemble

In models of allosteric regulation it is not unusual to assume a physical connection between a regulatory site, and a catalytic site that is influenced by binding at the regulatory site. Then, as an effector binds at its site, and initiates a new conformation, this is physically communicated to the nearby catalytic site. However, we now have examples where the connection between such sites is not rigid, and the different forms for an enzyme, with some sites properly formed, or unformed, could also be viewed as an ensemble.

A specific example of the dihydrofolate reductase from *E. coli* illustrates this. This enzyme binds two substrates, folate and NADPH, and when the enzyme is bound with either substrate, binding of the second substrate occurs with positive cooperativity.[127] Such results might normally be used to imply a direct physical link between the two sites, to help coordinate the conformations of the two. Since the protein structure is known, by using a program to generate a large number of different conformational states, it was observed that the region in the structure between the two binding sites never had any significant structure.[127] Therefore, binding of a ligand at one site, does not instantly stabilize the neighboring site.

However, it was shown that the protein exists as an ensemble of many conformations. A subset of such conformations will have the binding site for folate formed well enough for folate to bind and stabilize this subset. A somewhat different subset in the ensemble will have the binding site for NADPH formed well enough to favor the binding of NADPH. These two subsets will be different, but will also overlap, in that both subsets must include forms of the enzyme with both sites in the appropriate conformation. Therefore, the presence of either ligand will stabilize one subset. Although the second binding site is not automatically formed in this subset, it is more likely to occur, because many other conformations available to the free enzyme are not part of this subset, so that the correct conformation for the second ligand is more likely to occur. And this is observed as cooperative binding.

4.2.2 Evolution of Allosteric Enzymes

The key to making an enzyme allosteric is to have not only more than one folded state possible, since all enzymes do that, but to also have at least two separate conformations with different activities that can each be stabilized by ligand binding, or by covalent modification. As a simple example, in one conformation the active site is in a natural state, and accessible. The presence of substrate(s) able to bind here would further stabilize this conformation. To have an inactive conformation for this enzyme, it is only necessary to have an alternate conformation in which the active site is not properly formed or not accessible. However, it is also necessary to be able to stabilize this inactive conformation. If an additional binding site exists at which an inhibitor can bind and stabilize this enzymatically inactive conformation, then these properties produce an allosteric enzyme.

It appears possible that most enzymes may already be large enough that an additional binding site is potentially available, if the enzyme is flexible and can sample the appropriate conformational state in which this extra binding site becomes formed. Even if this conformation exists at a low frequency, the presence of ligands able to bind to it would then stabilize such a conformation, with the change in activity associated with such an enzymatic conformation. We actually have some limited examples of this (Table 4.2). For the six enzymes listed, scientists normally screened a library of thousands of potential compounds, and then characterized the best binding candidates. For each enzyme in this table, a crystal structure with the new inhibitor or activator clearly demonstrated that the compound bound at a site different than any of the previously established ligands of that enzyme.

The binding constants are generally strong enough that this novel site on each enzyme is as well defined as the catalytic site, and the other allosteric sites that were already known. No binding constant was obtained for the caspases since the new drug is a thiol-compound, designed to form a disulfide with a reactive cysteine on the enzyme. Since artificial drugs were used to detect each site, it is possible that these new sites may in fact have a normal physiological ligand that remains to be detected. It is equally possible that these are examples of existing sites that remained cryptic until a unique drug was found that matched them, since in these examples no mutations occurred to make these sites functional.

Table 4.2. Enzymes with newly discovered allosteric sites

Enzyme	New allosteric effector[a]	K_d or IC_{50} (nM)	Refs.
Caspase	DICA; FICA	nd	128
Fructose 1,6-bisphosphatase	Anilinoquinazoline	600	129
Glucokinase	RO-28-1675	~3,000	130
Glycogen phosphorylase a	CP-91149	130	131, 132
p38 MAP kinase	BIRB 796	0.1^b	133
Protein tyrosine phosphatase 1B	Compound 3	8,000	134

[a] Descriptions and structures of novel inhibitors are available in the original paper
[b] K_d value

In Chap. 8 we will also see an example of how a nonallosteric species of phosphofructokinase can be converted into the allosteric form by a single point mutation. Since almost all phosphofructokinases are well regulated by allosteric effectors, this implies that in this one case mutation of the ancestral allosteric enzyme led to a normal enzyme devoid of allosteric control. Therefore, reversing such an evolutionary mutation in the laboratory will "rediscover" the original allosteric binding site. And Chap. 4.3 presents uracil phosphoribosyltranferases, describing species in which the enzyme is allosteric, and others in which it is not, even though the enzymes have fairly high sequence identity. This suggests that occasional mutations during evolution may have made some allosteric sites cryptic, since for specific enzymes such sites have lost this regulatory function. But, what has been lost by mutation can also be recovered by mutation.

The importance of this hypothesis for the origin of allosteric enzymes is that it so naturally explains how almost one-third of all enzymes obtained this feature, and therefore makes it much more plausible that such allosteric properties were naturally and easily developed. Consider a population of microbes that have a metabolic pathway lacking an allosteric control point. If, for the potential regulatory enzyme, one or two mutations can unmask a cryptic binding site in a different conformation, then the cell with this emerging control will have a clear advantage by being more efficient. Its descendants should replace the previously normal cells, and with increased selective pressure the new allosteric control can become nicely adapted for optimum response within these cells' environment. No other mechanism for evolving so many allosteric enzymes is as simple or as plausible.

In later chapters we will also see examples where the original enzyme was increased in size by insertion of one or more modules/domains to provide additional binding sites, and these always have a regulatory function. For enzymes that are currently well defined, this mechanism provides additional binding sites to expand the control features for such enzymes, by making them sensitive to additional physiological effectors.

4.3 Uracil Phophoribosyltransferase: Different Regulatory Strategies for the Same Enzyme

A clear illustration of how regulation may be accomplished by different strategies is provided by examples of uracil phosphoribosyltransferase (UPRTase) from different organisms. The enzyme is designated as a salvage enzyme, since it permits the reutilization of the pyrimidine base uracil to form the nucleotide uridine monophosphate (UMP):

$$\text{Uracil} + \text{P-ribose-PP} \rightleftarrows \text{UMP} + \text{PP}_i$$

UPRTase in *B. caldolyticus* has normal kinetic properties, and shows no evidence of being regulated.[135] In most organisms studied, UPRTase is normally dimeric, and can associate to form higher oligomers such as tetramers or hexamers, which are the more active forms. These are then examples of dissociating regulatory enzymes, as depicted in Fig. 1.9.

The enzyme from *E. coli* has evolved to have an additional binding site for GTP, and when GTP is bound the enzyme is stabilized as a hexamer, and this species has a better K_m (fivefold lower), with no significant change in V_{max}.[136] This form of regulation defines the *E. coli* enzyme as *K*-type. The enzyme from *T. gondii* has a similar regulatory response, and is also *K*-type.[137] By comparison, the enzyme from *S. solfataricus* is always a tetramer, but is still activated by GTP, which results in raising k_{cat} by about 20-fold (Table 4.3).[138] This is therefore a *V*-type regulated enzyme.

Why is GTP an allosteric effector for an enzyme that produces UMP? In the synthesis of either DNA or RNA purine and pyrimidine nucleotides are required fairly equally. Whenever these microbes have abundant GTP (representing purines) they should make higher amounts of UTP and CTP, which come from UMP. Having all these nucleotides in comparable abundance then permits the cells to duplicate their genome, in preparation for cell division, and thereby increase their number. Therefore, the allosteric effect of binding GTP leads to increased activity, and this may result from a conformational change that makes binding more efficient, such as the *K*-type enzyme from *E. coli* and *T. gondii*, or that makes the total activity greater, such as the *V*-type enzyme from *S. solfataricus*.

Additional data for these enzymes in Table 4.4 show that the enzyme must be at least dimeric for activity. For the *B. caldolyticus* UPRTase this is achieved by having a high enough concentration of enzyme, though this would also be necessary simply to stabilize this enzyme since this organism is a thermophile. For the other microbes, association to higher oligomers is stabilized by the activator, GTP, as well as by one of the substrates, P-ribose-PP. The enzyme from *S. solfataricus* also has evolved a binding site for CTP, to act as an inhibitor. This provides a logical additional form of control. Because GTP and CTP are normally at similar concentrations, the binding constants in Table 4.4 for this enzyme show that it is more sensitive to inhibition by CTP. It is also evident that all these enzymes have very low catalytic rates, even when they have been activated.

It is also worth noting that any single specific metabolic step may or may not exist in different organisms, if they have an alternate means for providing the same end product. If it occurs, it may or may not be regulated, and if it is regulated it may be regulated by alternate strategies. While this may seem messy, or puzzling, it also emphasizes that organisms evolve by being opportunistic. Depending on the ecological niche for these different microbes, they may not have an equal need to control this specific metabolic step. While three of the bacteria in Table 4.3 have sophisticated regulation for uracil salvage, *B. caldolyticus* does not need such regulation. And mammals do not even have this enzyme. Mammals use an alternate pathway by recycling uridine instead of uracil.

An important point here is the variety that we see in how uracil is salvaged by different organisms. In addition, regulatory features are not essential, although they should

Table 4.3. Examples of uracil phosphoribosyltransferase in different organisms

Species	Regulation	Regulatory mode	Refs.
Bacillus caldolyticus	None	none	135
Escherichia coli	GTP reduces K_m 5 ×	*K*-type	136
Toxoplasma gondii	GTP reduces K_m 7 ×	*K*-type	137, 139
Sulfolobus solfataricus	GTP increases k_{cat} 20 ×	*V*-type	138

Table 4.4. Parameters for uracil phosphoribosyltransferase in different organisms

Species	Oligomer Change	Activator	K_d (μM) for GTP	K_d (μM) for CTP	k_{cat} (s^{-1})	Refs.
Bacillus caldolyticus	$\alpha \rightarrow \alpha_2$	Higher [E]	–	–	7.3	135
Escherichia coli	$\alpha_2 \rightarrow \alpha_6$	GTP	550	–	2	136
Toxoplasma gondii	$\alpha_2 \rightarrow \alpha_4$	GTP	465	–	0.7	137, 139
Sulfolobus solfataricus	Always α_4	GTP	85	5	2.5	138

be beneficial for those organisms that manifest them. The regulatory mechanism can then occur by different strategies, even though the same enzyme, UPRTase, and the same activator, GTP, are always involved.

The examples of UPRTase also suggest that regulation may have evolved separately and independently in organisms that had very similar forms of the same enzyme. This enzyme in *B. caldolyticus* shows no evidence of being regulated. This example may reflect, in the evolution of enzymes for salvaging uracil, the earliest version for this enzyme. Since other enzymes have different responses to the same regulatory effector, this supports the hypothesis of independent evolution to solve a similar need for regulation.

5

KINETICS OF ALLOSTERIC ENZYMES

Summary

The MWC model postulates that all subunits in an enzyme oligomer change conformation in a *concerted* fashion, while the KNF model assumes this change may be *sequential* for the different subunits. Most K-type enzymes have positive cooperativity; a limited number of regulatory enzymes have negative cooperativity. This controlled change in affinity makes the enzyme more sensitive to changes in substrate concentration (positive cooperativity), or dampens the enzyme's response to changes in substrate concentration (negative cooperativity). When multiple enzymes work in concert for some cellular process, and are regulated in concert by the same original effector, this may produce very high ultra-sensitivity.

5.1 Kinetics for Cooperative K-Type Enzymes

The majority of enzymes have a constant affinity for their substrates, and therefore always have a hyperbolic binding or kinetic curve, as defined in Chap. 3, and depicted in Fig. 5.1, left. The subset of allosteric enzymes that are regulated by a change in affinity may have either of the two types of kinetic patterns shown for positive cooperativity or for negative cooperativity (Fig. 5.1). Both of these cooperative patterns are the natural consequence of a change in the enzyme's affinity with changing substrate concentration.

I described earlier how the linear response to a ligand at constant affinity goes over a range of almost 100-fold in the concentration of the substrate or ligand being bound, and mentioned that the *2-log rule* is an easy way to remember this. To define the most linear part of such a curve, Koshland et al.[24] introduced the term for *enzyme sensitivity*, R_S, defined as the ratio of [S] that will produce velocities at 0.9 and 0.1 of V_{max}. For an enzyme with constant affinity this requires a change in concentration of 81-fold. Although allosteric enzymes have a very poor binding response at low [S], when they do begin to accelerate their response, the linear part of the velocity curve is now steeper

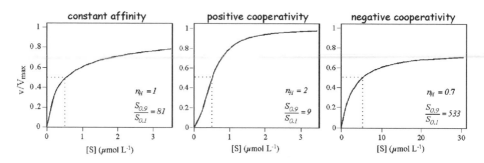

Fig. 5.1. Types of sensitivity of enzymes with allosteric features. Normal sensitivity implies constant affinity ($n_H = 1.0$); for this curve $K_m = 0.5$ μM. Positive cooperativity means greater sensitivity to changes in [S], ($n_H > 1.0$); for this curve $K_{0.5} = 0.5$ μM. Negative cooperativity means a dampened response to [S], ($n_H < 1.0$); for this curve $K_{0.5} = 5$ μM. Note the expanded scale on the abscissa for negative cooperativity

than for a normal enzyme, and in the example of Fig. 5.1, the same change in velocity now requires only a ninefold increase in [S]. The extent of cooperativity modeled in Fig. 5.1 is for positive cooperativity with an n_H of 2, which is normal for allosteric enzymes. When an enzyme or carrier protein has a higher n_H the kinetic curve becomes steeper.

We will see in Chap. 6 that hemoglobin, with an n_H of 3.2, has an $S_{0.9}/S_{0.1} = 4$. We will see that this degree of cooperativity is essential for hemoglobin, since it operates between the lungs and the tissues, with a change in the surrounding oxygen that is close to fourfold, thereby allowing hemoglobin to load up with oxygen almost completely in the lungs, and unload up to 70% of the bound oxygen in the tissues.

Enzymes or receptors that have constant affinity will always require an 81-fold change in substrate (or other ligand) concentration to go from 10% saturation to 90% saturation. Therefore, positive cooperativity is a mechanism to make the protein more sensitive to changes in [S], while negative cooperativity dampens this response. The full range of observed cooperativity is illustrated in Fig. 5.2, where it is evident that with negative cooperativity the substrate concentration ratio approaches 10,000, while with the highest degree of positive cooperativity this ratio is reduced to about 1.1. This range defines the currently known extent to which negative and positive cooperativity can make enzymes more, or less responsive in how they bind their normal substrates, or other physiological effectors.

To understand this range for cooperativity, let us consider it in a physiological context. The majority of metabolites in central metabolism, such as simple sugars, amino acids, nucleosides and nucleotides may vary by up to tenfold in the blood of modern humans as they alternate between daytime meals and overnight fasting. But within our tissues these changes are not as great, since the liver is designed to remove excess lipids, sugars, fats, and amino acids from blood. These are then metabolized and stored as glycogen or fat. During fasting fats and sugars are then released into blood for use by the tissues, and new nucleotides are easily synthesized. Only the essential amino acids can not be internally replaced, but must come from continued nutrition.

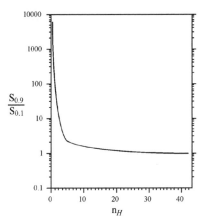

Fig. 5.2. The range of observed cooperativity, and the corresponding substrate concentration ratios to go from 10% of saturation to 90% of saturation. n_H values >6 require cascades with several enzymes

But for humans, before they had developed a steady food supply, and for all undomesticated animals, periods of starvation were or still are normal. Animals must be able to survive winter or any other condition that produces a dietary insufficiency. Such periods of starvation may produce changes in normal metabolites, and such survival was assured by the early evolution of appropriate regulatory controls for many enzymes.

For the range depicted in Fig. 5.2 enzymes with significant negative cooperativity ($n_H \leq 0.5$) have no meaningful change in activity, even if the substrate concentration changes by 100-fold, which is near the maximum of what is normally possible in vivo. Instead of being controlled by changes in [S], such enzymes would have a fairly steady rate in a metabolic context where large changes in the concentration of the substrate occurred. An example of such a need is observed with CTP synthetase.[140] Even during periods of caloric restriction, cells must continue to make messenger RNAs that enable some level of continued protein synthesis. At least a moderate level of CTP synthesis is then always required, even when the concentration of the substrate glutamine may become limiting. We will consider this enzyme later, and see that this is accomplished by having a high affinity, so that activity is adequate even at low [glutamine], but the activity does not increase normally with higher [glutamine] due to the dampened saturation of negative cooperativity.

For enzymes with significant positive cooperativity ($n_H = 2.5$) a fivefold increase in the substrate concentration can increase the activity from 10% of V_{max} to 90% of V_{max}. But it is more likely that even when there is no change in the substrate concentration for some control enzyme, it will be necessary for it to alter its rate. A simple example is glycolysis, for which phosphofructokinase is an important control point, and this enzyme will be described in Chap. 8. Under conditions where the substrate for this enzyme, fructose-6-P, is not changing in concentration, an organism may perceive an emergency requiring quick action. This will be accompanied by release of epinephrine (adrenaline) and the hormone signal results in changing the conformation of this enzyme leading to a

Table 5.1. Assumptions/conditions for the MWC Model[a]

1. Allosteric proteins are oligomers
2. For each ligand able to bind to a subunit, there is only one binding site per subunit
3. The conformation of each subunit is constrained by association with other subunits
4. At least two conformational states are accessible to allosteric oligomers. These states differ by the distribution and/or energy of intersubunit bonds, and therefore by the conformational constraints imposed upon the subunits
5. Therefore the affinity of one or more of the stereospecific sites toward the corresponding ligand is altered when a transition occurs from one to the other state
6. When the enzyme oligomer goes from one state to the other state, its molecular symmetry is conserved

[a]These initial assumptions have not proven to be invariant, and exceptions to three of the assumptions (1, 2, and 6) will be described further on. Glucokinase is monomeric; phosphofructokinase has three sites per subunit for ATP; many allosteric enzymes do not conserve symmetry

decreased K_m for fructose-6-P and therefore increased activity (muscle enzyme) and also an increased K_m for fructose-6-P and therefore decreased activity (liver enzyme).

5.1.1 The MWC Model for Positive Cooperativity

The first model to define the kinetic analysis of enzymes with positive cooperativity was published by Monod, Wyman, and Changeux (the MWC model).[23] Although the assumptions these authors made have not proven to be invariant (some exceptions are listed in the footnote to Table 5.1), it is worthwhile to start with the initial MWC model and then consider extensions and variations. Additional notation defines subunits in the active R and less active T conformations with $0-n$ ligands bound as $R_0, R_1, R_2, R_3,...R_n$, and $T_0, T_1, T_2, T_3,...T_n$. These lead to the following binding equilibria for any ligand, where L may represent a substrate or effector molecule:

$$R_0 \leftrightarrow T_0$$

$$R_0 + L \leftrightarrow R_1 \qquad\qquad T_0 + L \leftrightarrow T_1$$
$$R_1 + L \leftrightarrow R_2 \qquad\qquad T_1 + L \leftrightarrow T_2$$
$$\cdots\cdots\cdots \qquad\qquad \cdots\cdots\cdots$$
$$R_{n-1} + L \leftrightarrow R_n \qquad\qquad T_{n-1} + L \leftrightarrow T_n$$

These equations are then refined by including the probability of L binding to the first subunit of the empty oligomer, up to its binding to the nth subunit. When L encounters the free oligomer, there are n possible binding sites, and after $n-1$ sites are filled, there is but $1/n$ of the binding sites remaining.

$$T_0 = LR_0$$

$$R_1 = R_0 n \frac{L}{K_R} \qquad\qquad T_1 = T_0 n \frac{L}{K_T}$$

$$R_2 = R_1 \frac{n-1}{2} \frac{L}{K_R} \qquad\qquad T_2 = T_1 \frac{n-1}{2} \frac{L}{K_T}$$

$$\cdots\cdots\cdots \qquad\qquad\qquad \cdots\cdots\cdots$$

$$R_n = R_{n-1} \frac{1}{n} \frac{1}{K_R} \qquad\qquad T_n = T_{n-1} \frac{1}{n} \frac{1}{K_T}$$

In these equations K_R and K_T represent the affinity of the R and T conformations for the ligand. The two conformations do not have equal activity because the affinity K_T is normally 20–100-fold poorer than K_R, and this ratio is represented as c:

$$c = \frac{K_R}{K_T}. \tag{5.1}$$

The fraction of the enzymes in the more active state, $\overline{R}$, can then be logically defined as

$$\overline{R} = \frac{R_0 + R_1 + R_2 + \cdots + R_n}{(R_0 + R_1 + R_2 + \cdots + R_n) + (T_0 + T_1 + T_2 + \cdots + T_n)}. \tag{5.2}$$

The other important parameter is the extent to which all the binding sites are saturated, $\overline{Y}_L$. This term must include the probability factor for binding to each site in an oligomer.

$$\overline{Y}_L = \frac{(R_1 + 2R_2 + \cdots + nR_n) + (T_1 + 2T_2 + \cdots + nT_n)}{n[(R_0 + R_1 + R_2 + \cdots R_n) + (T_0 + T_1 + T_2 + \cdots + T_n)]}. \tag{5.3}$$

To simplify (5.2) and (5.3) for the binding of substrate, they substituted the terms:

$$\alpha = \frac{[S]}{K_R} \tag{5.4}$$

in which the substrate concentration is normalized by the enzyme's affinity for the substrate, and an *allosteric constant*, L,[*] that defines the equilibrium between inactive and active conformations:

$$L = \frac{[E_T]}{[E_R]} \tag{5.5}$$

to provide the final equations:

$$\overline{R} = \frac{(1+\alpha)^n}{L(1+c\alpha)^n + (1+\alpha)^n}, \tag{5.6}$$

$$\overline{Y}_L = \frac{Lc\alpha(1+c\alpha)^{n-1} + \alpha(1+\alpha)^{n-1}}{L(1+c\alpha)^n + (1+\alpha)^n}. \tag{5.7}$$

When the affinity of the T conformer is very poor, K_T will be very large, so that c becomes very small, and (5.7) simplifies to:

$$\overline{Y}_L = \frac{\alpha(1+\alpha)^{n-1}}{L + (1+\alpha)^n}. \tag{5.8}$$

---

[*]To distinguish this allosteric constant, defined as L in the MWC model, from any generic ligand, L, the allosteric constant will always be in italic.

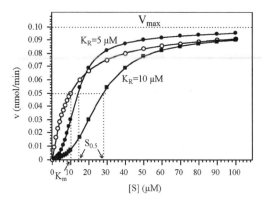

Fig. 5.3. Examples of positive cooperativity. *Open circle*, the curve from Fig. 3.2 for constant affinity, with $K_m = 10 \, \mu M$; *filled square*, with $K_R = 10 \, \mu M$ and $K_T = 1{,}000 \, \mu M$; *filled circle*, with $K_R = 5 \, \mu M$ and $K_T = 500 \, \mu M$

Alternatively, even if there are two conformations, should they have the same affinity for the substrate, S, (then $c = 1$), (5.7) simplifies to the normal Michaelis–Menten expression for which the affinity is constant:

$$\overline{Y}_S = \frac{\alpha}{1+\alpha} = \frac{[S]}{K_R + [S]}. \qquad (5.9)$$

It is important to note that the saturation function, $\overline{Y}_S$, is comparable to v/V_{max}. We can then use (5.9) to create a kinetic plot for an enzyme that is allosteric, and for a specific example we will assume that $L = 100$. This means that the starting distribution between the two conformers is almost completely in the T form. When the enzyme is almost completely in the T conformation, the activity at low $[S]$ is very low because the T conformer binds substrate very poorly (Fig. 5.3). For this curve in Fig. 5.3c has been assumed to be 0.01, so that the affinity of K_R will be $100 \times K_T$. Only after the substrate concentration has increased enough to pull the R/T distribution more toward R (see Fig. 1.4) does the activity become significantly greater. It is this feature that produces the sigmoid curve that is diagnostic of allosteric enzymes with positive cooperativity. If we then also have K_R be the same as the K_m for the data set plotted in Fig. 3.1, we can produce the examples of a kinetic curve for positive cooperativity shown in Fig. 5.3.

In modeling allosteric kinetic curves we have two important parameters that may be varied: L, the ratio of $[E]_T/[E]_R$ in the absence of ligands; and c, the ratio of their affinities, K_T/K_R. The actual affinity, K_R, of the more active form does not need to be specified, since the format for plotting the data is a plot of v vs. α, where α is the substrate concentration normalized by the affinity for that substrate (as examples see Figs. 5.4–5.6). All parameters for the two allosteric curves (Fig. 5.3) are the same, except that for the curve at the right, $K_R = 10 \, \mu M$, equal to the K_m for the hyperbolic curve, and for the middle curve $K_R = 5 \, \mu M$. This should emphasize the initial influence of the T form at low substrate concentrations, since the curve for the enzyme with the better affinity shows even a poorer activity at low $[S]$ compared to the standard curve for constant affinity.

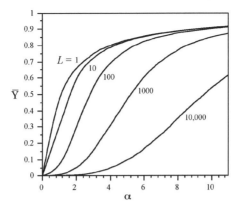

Fig. 5.4. Theoretical curves for $\overline{Y}$, as a function of L. *Curves* are calculated for a tetramer and with $c = 0$

Also shown in Fig. 5.3 is that one cannot visually extract the two affinity constants for an allosteric curve in the same manner used for a hyperbolic curve. For the two allosteric curves, the direct correspondence of the substrate concentration equal to $1/2V_{max}$ yields an average affinity for the allosteric enzyme population, normally denoted by $K_{0.5}$. For the two curves, this value is significantly greater than K_R, but – it is also not just an average of K_T and K_R, because for each case K_T was assumed to be $100 \times K_R$, so that the average of these two values would be far greater than the observed $K_{0.5}$. With appropriate software one can now do curve fitting to obtain a best estimate of the two different affinity constants for an experimental allosteric kinetic curve.

To explain the great change in activity for such allosteric curves with increasing [S], the term *cooperativity* was introduced. This was intended to describe how the binding of a substrate molecule to only one subunit in a tetramer could stabilize the complete oligomer in the R conformation, so that binding to the additional subunits in the same favorable conformation would be facilitated, as if the subunits cooperated with each other via their physical interactions. This feature is inherent in the MWC model with the assumption that all subunits in any oligomer always have the same conformation.

5.1.2 Influence of the Key Parameters on the Extent of Cooperativity

5.1.2.1 Homotropic Effects

The two principle parameters that define the extent of cooperativity are the constants L ($= [E_T]/[E_R]$) and c ($=K_T/K_R$). As L becomes greater, more of the free enzyme must be in the T conformation, and this means that the activity at low [S] will be very low. This is illustrated in Fig. 5.4, where the abscissa, α, represents the normalized substrate concentration ([S]/K_m), so that curves for any enzyme may be directly compared. It is evident that at large values of L the curves are dramatically sigmoid, since the initial activity is very poor. As L approaches 1, the two conformers become more equal in

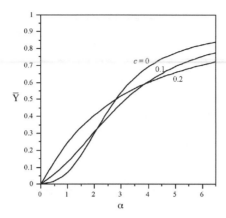

Fig. 5.5. Theoretical curves for $\overline{Y}$, as a function of c. *Curves* are calculated for a tetramer and with $L = 100$

abundance, and the curve approaches a hyperbolic form, indicative of a constant affinity. Since the effect being measured shows the influence of increasing [S], normalized as α, on the ability of the enzyme to bind *more of the same ligand* (the substrate in this case), this is called *homotropic cooperativity*.

When c is zero, this denotes that the T conformer has no affinity for the substrate (or other ligand). The effect of c for this situation provides the most dramatic cooperativity, as illustrated in Fig. 5.5. As c becomes modestly greater the curves steadily approach a hyperbolic form. The curve for $c = 0.2$ illustrates that the difference in the affinity of the R and T conformers is now not great enough for this curve to be distinguished from a completely hyperbolic curve for an enzyme with constant affinity. By comparing Figs. 5.4 and 5.5 it is evident that L is the more important constant.

5.1.2.2 Heterotropic Effects

Since allosteric enzymes are able to bind effectors as well as substrates, the influence of one type of ligand on the binding of the others is described as *heterotropic cooperativity*. The MWC model proposes that heterotropic effects only influence L, the equilibrium between the two conformers, and that it has no effect on c. This permits the more general (5.7) to be simplified as (5.10).

$$\overline{Y}_S = \frac{\alpha(1+\alpha)^{n-1}}{L' + (1+\alpha)^n},\qquad(5.10)$$

where L' is an apparent allosteric constant, defined as:

$$L' = \frac{\sum_0^n T_1}{\sum_0^n R_A}\qquad(5.11)$$

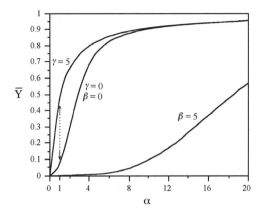

Fig. 5.6. Theoretical curves for $\overline{Y}$, as a function of β or γ. *Curves* are calculated for a tetramer with $L = 100$

in which the T conformers bind the inhibitor, while the R conformers bind with activators, consistent with the diagram in Fig. 1.4. Replacing L' in (5.11) it becomes

$$L' = L \frac{(1+\beta)^n}{(1+\gamma)^n}. \tag{5.12}$$

Introducing $\beta = [I]/K_I$ and $\gamma = [A]/K_A$, for the normalized concentrations of an inhibitor or an activator, (5.10) becomes

$$\overline{Y}_S = \frac{\alpha(1+\alpha)^{n-1}}{L\dfrac{(1+\beta)^n}{(1+\gamma)^n} + (1+\alpha)^n}. \tag{5.13}$$

This equation is intentionally kept somewhat simpler, by the assumption that inhibitors bind only to the T conformer, and activators only to the R conformer (see Fig. 1.4). Simulations for the effects of these constants are shown in Fig. 5.6.

The central curve in Fig. 5.6 is the control enzyme. As discussed in Chap. 3, the K_m of the enzyme is normally comparable to [S], which is the same as $\alpha = 1$ for the figure. At this substrate concentration the allosteric enzyme has modest activity, but this can easily be increased fivefold or more by an activator present at a reasonable concentration. One might interpret Fig. 5.6 to show a more pronounced effect by the inhibitor, when it is at a comparable normalized concentration to the activator, since at an inhibitor concentration of $5 \times K_i$ the inhibition appears more dramatic even at substrate concentrations greater than $10 \times K_m$. This visual impression is due to the scaling effect of such figures, since inhibition can shift that curve to the right over large ranges of substrate concentrations, but, at physiological concentrations of substrate ($\alpha = 1$), the actual activity can only go from about $0.1V_{max}$ of the control curve ($\gamma = 0$, $\beta = 0$) to almost 0. By comparison, at that substrate concentration an activator, at a concentration $5 \times K_a$ increases the activity from about $0.1V_{max}$ to almost $0.5V_{max}$.

5.2 RNA Riboswitches also Show Allosteric Binding

With our present knowledge we understand enzymes to overwhelmingly be proteins, though we are aware of a few essential RNA ribozyme activities. Bacteria continue to use many RNA molecules as *riboswitches*, which upon binding some specific metabolic ligand alter their structure and are then able to bind to a DNA regulatory site to influence the expression of genes coding for enzymes that specifically metabolize the ligand that activates that particular riboswitch. Most such riboswitches appear to have a single ligand binding site.[141]

As evidence that allosteric mechanisms function in the RNA world, an important study by Mandal et al.[141] demonstrated that a specific RNA riboswitch from *Vibrio cholerae* has two aptamers with binding sites for glycine, VC I and VC II. The genes controlled by this riboswitch express enzymes for the catabolism of glycine as an energy source, and ideally should only be turned on when glycine is abundant, to permit its normal use for protein synthesis. Protein enzymes have cooperativity due to the fact that they have more than one binding site within one oligomer, even though this is in most cases proportional to a stoichiometry of one per subunit. While RNAs do not normally form oligomers, an allosteric response is possible with this riboswitch by virtue of the fact that it contains two binding sites on a single RNA chain. Binding of a glycine at the first of these sites then induces a conformational change in the RNA leading to formation of the adjacent site to enable its binding of glycine (Fig. 5.7A), which explains the classic allosteric binding curve shown in Fig. 5.7B.

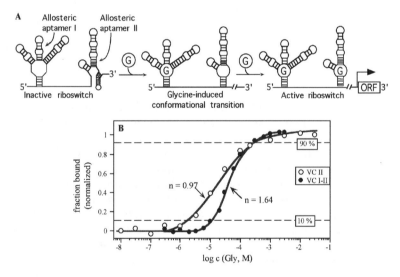

Fig. 5.7. An RNA riboswitch with cooperative binding for glycine. (**A**) Schematic illustration of the RNA with two aptamer sites that are able to bind glycine. In the free state, aptamer II is not yet properly structured to bind glycine. After one molecule of glycine has bound to aptamer I, a conformational change occurs to organize the aptamer II site. From M. Famulok, Science **306**, 233 (2004). Reprinted with permission from AAAS. (**B**) Binding curves for glycine with the normal VC I–II containing both sites, and with the altered VC II, containing only one site. From M. Mandal et al., Science **306**, 275–279 (2004). Reprinted with permission from AAAS

These authors demonstrated that the allosteric kinetics represented true site–site interaction by making a modified version of the riboswitch in which site I has been deleted. As shown in Fig. 5.7B, the modified RNA (VC II) now has a simple hyperbolic binding curve, consistent with a single site at a constant affinity. The allosteric response is ideal for a switch, since it greatly narrows the concentration range over which the riboswitch becomes fully active. This work with ribozymes supports the hypothesis that such regulatory mechanisms evolved very early in the emergence of living cells, since the benefits thereby provided would have had a selective advantage within the context appropriate for such progenitor forms.

5.3 Negative Cooperativity

To help define the response sensitivity of more complex life processes to a single quantifiable stimulus molecule, the concept of *ultrasensitivity* was introduced by Koshland.[142] Typical response curves are shown in Fig. 5.8 for normal sensitivity, ultrasensitivity, and subsensitivity. These correspond to the well-defined kinetic plots for constant affinity, as shown by most simple enzymes; to positive cooperativity, as shown in earlier figures in this chapter; and to negative cooperativity, which will be discussed further on. To help understand the influence of cooperativity on an enzyme, the enzymes at different n_H in Fig. 5.8, are plotted together in Fig. 5.9.

The earliest enzymes that showed a change in affinity had kinetic curves with the sigmoid features evident in Figs. 5.3–5.6, and the term cooperativity became used to suggest the interaction between subunits to account for the conformational transformation. A few enzymes, however, had another type of kinetic curve, as shown in Fig. 5.8A, right. Such kinetic data, when plotted in the Lineweaver–Burk format (Fig. 5.8B, right) produce a downward curvature. In contrast, enzymes with positive cooperativity that normally have a sigmoid curve in linear plots (Fig. 5.8A, center), have an upward curvature in the Lineweaver–Burk format (Fig. 5.8B, center). Since it had become evident that these were both examples of cooperativity, the curve that bends upward was defined as *positive cooperativity*, and the downward curve as *negative cooperativity*.

5.3.1 Different Mechanisms Produce Negative Cooperativity

Positive cooperativity occurs only when a conformational change occurs upon binding some ligand, which stabilizes the more active conformation for the whole oligomer. The kinetics indicative of negative cooperativity may be observed under different conditions, or due to different mechanisms:

1. An asymmetric oligomer, with subunits in different conformations that have different affinities.
2. Half-of-the-sites activity, because the substrate is large enough that in binding to one subunit it sterically occludes binding to the adjacent subunit.
3. Half-of-the-sites activity, because the substrate on binding to one subunit produces a conformational change in the adjacent subunit that makes binding very difficult or impossible.

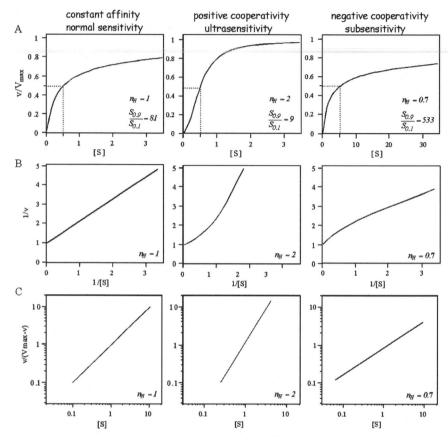

Fig. 5.8. Ultrasensitivity and subsensitivity, compared to constant affinity plots; as depicted (**A**) in normal linear plots, (**B**) in Lineweaver–Burk plots, and (**C**) in Hill plots. The same kinetic values were used as in Fig. 5.1. For (**A**), right, note the change in scale for the abscissa. Normal sensitivity implies constant affinity ($n_H = 1.0$); for this curve $K_m = 0.5\ \mu M$. Ultrasensitivity implies positive cooperativity ($n_H > 1.0$); for this curve $K_{0.5} = 0.5\ \mu M$. Subsensitivity implies negative cooperativity ($n_H < 1.0$); for this curve $K_{0.5} = 5\ \mu M$

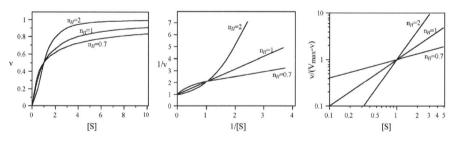

Fig. 5.9. Direct comparison of enzymes with ultrasensitivity or subsensitivity to enzymes with constant affinity. The data from Fig. 5.8 were used

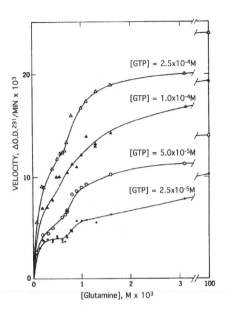

Fig. 5.10. Negative cooperativity for the binding of glutamine by CTP synthetase. [From Levitzki & Koshland, Proc. Natl. Acad. Sci. USA, **62**, 1121–1128 (1969)]

Condition #1 is the most frequent mechanism for producing negative cooperativity, and has been supported by crystal structures of several enzymes, where oligomers with subunits of different conformations have been observed.[142]

Experimental kinetic data for CTP synthetase are shown in Fig. 5.10.[143] The enzyme shows kinetic curves that are classic for negative cooperativity vs. the substrate glutamine, which serves as the amino group donor in the reaction by which the pyrimidine nucleotide UTP is converted to CTP. The purine nucleotide GTP acts as an activator, since the activity increases at higher concentrations of GTP. The lowest curve in this figure has the appearance of being two separate hyperbolic curves added together. The first half of this curve appears to be reaching a plateau below 1 mM glutamine, representing the V_{max} of the high affinity subunit. A second, higher plateau is suggested, but occurs at concentrations of the substrate above those used in this study, and represents the V_{max} of the second subunit, which has much poorer affinity. As the activator GTP was increased, not only does the total activity increase, but the appearance of two separate hyperbolic curves is less obvious, and at the highest GTP concentration the data points almost define a single hyperbolic curve. For this last curve, negative cooperativity is just barely evident. The activator, GTP, has nearly saturated the enzyme, leading to almost all the subunits being in the high affinity conformation, and this should then produce a hyperbolic kinetic curve.

As was briefly mentioned at the beginning of this chapter, the important kinetic features in this figure show that the high affinity subunit has a K_m near 80 μM, while the low affinity subunit has a K_m near 700 μM. The available concentration of glutamine will rarely be below 50 µM, but may normally be in the low millimolar range. Then the benefit of this form of cooperativity is that CTP synthesis will be reasonably active even at quite low concentrations of glutamine, but will not increase to too high a level when glutamine concentrations are nearer their normal levels. Because CTP synthetase is an allosteric enzyme, this form of cooperativity toward glutamine allows the enzyme to be insensitive to concentration changes for this particular substrate, whose concentration fluctuates with its role in energy metabolism. It then permits the enzyme to be sensitive to other effectors that more accurately signal the cell's need for CTP.

5.3.1.1 Half-of-the-Sites Activity

Half-of-the-sites represent an extreme form of negative cooperativity. If we consider simply a dimer, for standard negative cooperativity normally one subunit in the dimer has a lower affinity than the other subunit, but both have the potential for full activity if [S] could be made high enough. This is depicted by the kinetic curves in Fig. 5.10. With half-of-the-sites, the second subunit is almost unable to bind the same ligand. This may result from steric effects when the substrate is large and blocks access to the catalytic site of the adjacent subunit, or may be caused by a conformation that occurs which has very poor binding.

This has been demonstrated with binding studies, where it has been shown that the CTP synthetase dimer could bind two molecules of the substrate glutamine, but only one molecule of the modifying reagent deoxo-norleucine (DON), a competitive inhibitor of glutamine.[140] Therefore CTP synthetase exhibits normal negative cooperativity with glutamine (illustrated in Fig. 5.10), and half-of-the-sites cooperativity with DON. Similarly, the tetrameric glyceraldehyde 3-phosphate dehydrogenase became completely inactivated upon binding two molecules of p-chloromercuribenzoate.[144]

5.3.2 Improper Enzyme Samples May Produce Erroneous Negative Cooperativity

There are also situations that may incorrectly lead to the interpretation of negative cooperativity:

1. Mixed population of different enzymes that have overlapping activities, but very different affinities for the same substrate.
2. Mixed population of the same enzyme where a subset has become partially denatured and therefore has poorer affinity for the same substrate.

For these latter examples, the native enzyme being studied may have a normal, constant affinity for its substrate(s). But if the enzyme sample contains other proteins that have a weaker activity with the same substrate, this extra activity will be observed at very high substrate concentration. The V_{max} will then be the sum of the activity of the principal enzyme, plus the activities of the one or more extra enzymes that are contaminants in the assay sample. The affinity, $K_{0.5}$, will then reflect the average position of all enzyme molecules that produces an activity equal to 50% of V_{max}.

With the many improved techniques for purifying enzymes, mixed enzyme samples are seldom a problem any more. However, condition #2 above may easily occur with a pure preparation of enzyme, if some feature of the storage buffer or assay buffer leads to a portion of the enzyme mixture becoming partially denatured. Such denatured enzyme molecules may still have some reduced activity, or poorer affinity, so that these denatured molecules will contribute to the observed activity at very high [S]. Naturally, if almost all the enzyme molecules become denatured, the population will again be homogeneous and produce a normal binding/activity curve, albeit with a constant but very high K_m.

When an enzyme sample is demonstrably pure, it may be difficult to discern that any denaturation has occurred. In some cases, such denaturation is reversible, so that one may assay a truly native enzyme sample, as well as a partially denatured sample. This was demonstrated with the enzyme purine nucleoside phosphorylase, which becomes somewhat less active upon oxidation, and this was achieved by bubbling pure oxygen gas into the enzyme sample.[145] Such oxidized enzyme then produced kinetics that were typical for negative cooperativity. When the same enzyme sample was then treated with dithionite to remove the oxygen, the observed kinetic activity showed only a standard hyperbolic curve, consistent with a constant affinity of a homogeneous enzyme population.

5.3.3 Morpheeins Display Negative Cooperativity

Enzymes that may alternate between alternative olgomeric states with distinctly different kinetic features were briefly described earlier (Chap. 1.2.3.3). Such enzymes may exist as a steady-state mixture of the two forms with different kinetic features. Therefore, the equation to describe the velocity of such a mixture is a double hyperbolic equation (5.14), and the sum of the two terms defines a kinetic curve such as those in Fig. 5.10.

$$v = \frac{V_{max_1}[S]}{K_{m_1} + [S]} + \frac{V_{max_2}[S]}{K_{m_2} + [S]}. \tag{5.14}$$

5.3.4 Comparison of the MWC and KNF Models

One of the distinctive features of the KNF model is that it is the only model that predicts, and is able to calculate the factors for negative cooperativity. To see what makes this possible, let us consider a simple dimer of two identical subunits and each with a binding site for L. This leads to the two models shown in Fig. 5.11. An immediate feature evident for negative cooperativity is that these enzymes never reach saturation in binding substrates, or effectors, and never reach V_{max} (Fig. 5.8A, right).

The equations for the MWC model were described earlier, and we will now examine the KNF model, where the equilibria for binding in a sequential fashion are:

$$E + L \underset{\longleftarrow}{\overset{K_1'}{\longrightarrow}} EL, \tag{5.15}$$

$$EL + L \underset{\longleftarrow}{\overset{K_2'}{\longrightarrow}} EL_2. \tag{5.16}$$

K_1' and K_2' are the statistically corrected dissociation constants, and these are related to the thermodynamic dissociation constant by the equation

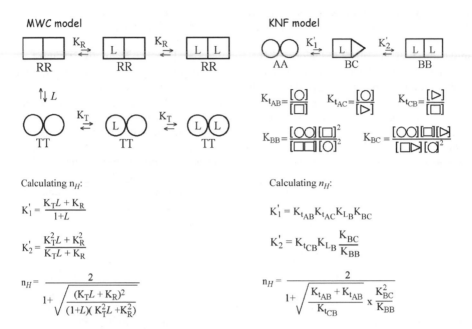

Fig. 5.11. Comparison of the MWC and KNF models for a dimer. R and T conformations have been defined; the KNF model has more conformations, and employs A, B, and C for the forms indicated. Additional dissociation constants are defined for the KNF model

$$K_i = [(n\mathrm{M} - i + 1)/i]K_1',$$
(5.17)

where K_i is the constant for the ith member of an enzyme containing a total of $n\mathrm{M}$ sites, and K_i' is the statistically corrected constant. Then for the dimer,

$$K_1 = 2K_1'$$

and

$$K_2 = (1/2)K_2'.$$

Concentrations of the bound forms EL and EL2 are given by

$$[EL] = 2([E][L])/K_1',$$
(5.18)

$$[EL_2] = \frac{([EL][L])}{2K_2'} = \frac{[E][L]^2}{K_1'K_2'},$$
(5.19)

$$\overline{Y} = N_L/n\mathrm{M} = \frac{1}{2}([EL]+2[EL_2])/[E_{total}] \qquad (5.20)$$

or

$$\overline{Y} = \frac{1/2([EL]+2[EL_2])}{([E]+[EL]+[EL_2])}, \qquad (5.21)$$

where n_M is the number of sites and N_L the number of molecules of L bound for any concentration $[L]$. Inserting (5.18) and (5.19) into (5.21) leads to

$$\overline{Y} = \frac{[L]+[L]^2/K_2'}{K_1'+2[L]+[L]^2/K_2'}, \qquad (5.22)$$

and then

$$\frac{\overline{Y}}{1-\overline{Y}} = \frac{[L]+[L]^2/K_2'}{K_1'+[L]}. \qquad (5.23)$$

From (5.23) it follows that the concentration of L that gives 50% saturation is

$$L_{0.5} = (K_1'K_2')^{1/2}. \qquad (5.24)$$

Applying this value to the Hill equation, we obtain

$$n_H = \frac{2}{1+(K_2'/K_1')}. \qquad (5.25)$$

We can now test the two theories, by their equations, for expression of the Hill coefficient for a dimer, and thereby quantitate the extent of cooperativity (i.e., n_H) under the possible variations for the two affinity constants. The two models and the respective equations are summarized in Fig. 5.11. The following combinations of affinity constants are possible:

1. $K_1' = K_2'$
2. $K_1' > K_2'$
3. $K_1' < K_2'$

Clearly condition #1 is the standard condition for all enzymes with no cooperativity. Condition 2 applies to those enzymes with positive cooperativity. Condition 3 describes the case for enzymes with negative cooperativity.

Both models will calculate n_H to be 1.0 for condition 1. Inspection of the MWC equation for n_H (Fig. 5.11) shows that the term whose square root will be calculated must be ≤ 1.0 for all possible values of L, K_R and K_T. Therefore, as this term approaches 1.0, n_H approaches 1.0, and as this term approaches 0, n_H approaches 2.0. The equations of the MWC model therefore do not permit n_H values less than 1.0, and therefore cannot express negative cooperativity.

For the more complex KNF model there are values for the different constants that permit n_H to vary from 0 to 2.0. This demonstrates the greater range and flexibility of the KNF model, since it can describe both positive and negative cooperativity. An important feature of both models is that n_H cannot exceed the number of subunits.[†] The KNF model can be expanded to tetramers or other oligomers, with the introduction of additional binding constants for the additional subunits. Since symmetry of the subunits is not required in the KNF model, then there is also no necessary hierarchy in the binding constants. That is $K_1' > K_2' = K_3' < K_4'$, and other combinations are possible.

In a practical sense, the general equation for the KNF sequential model is more versatile, since it covers the condition where change is concerted (i.e., the MWC model), as well as the conditions where different conformations coexist as subunits change in a sequential pattern. But, this general equation is certainly more complex, as shown in (5.28) with the following definitions: $K_t = [R]/[T]$, K_S = affinity for substrate, and

$$K_{TR} = \frac{[TR][T]}{[TT][R]}, \tag{5.26}$$

$$K_{RR} = \frac{[RR][T][T]}{[TT][R][R]}, \tag{5.27}$$

$$\overline{Y} = \frac{(K_{TR}^2 K_S K_t[S]) + (K_{TR}^4 + 2K_{TR}^2 K_{RR} K_S K_t[S])^2 + (3K_{TR}^2 K_{RR}^2 K_S K_t[S])^3 + (K_{RR}^4 K_S K_t[S])^4}{1 + (4K_{TR}^2 K_S K_t[S]) + (2K_{TR}^4 + 4K_{TR}^2 K_{RR} K_S K_t[S])^2 + (4K_{TR}^2 K_{RR}^2 K_S K_t[S])^3 + (K_{RR}^4 K_S K_t[S])^4}$$

$$\tag{5.28}$$

In (5.28) K_S is a normal affinity constant comparable to K_m, and K_t is similar to the allosteric constant, L, of the MWC model. The greater accuracy and complexity of this equation are due to the extra terms for the interaction between subunits in the two conformations, K_{TR} and K_{RR}. With the limited availability of computers in the 1960s, it would have been quite laborious to use (5.28) for any extensive curve fitting and

---

[†]When cooperativity results only from the slow kinetics for interconverting conformations, the ratio of n_H to the number of subunits need not remain ≤ 1. We will see an example in Chap. 10 with the monomeric glucokinase for which n_H = 1.5–2.3.

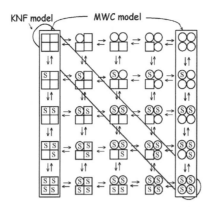

Fig. 5.12. Possible conformational changes for the four subunits in a tetramer. The MWC model is represented by the two end columns; the KNF model by the diagonal from the *top left* to the *bottom right*

Koshland et al. discussed this difficulty. The complexity of (5.28) provides for a more sophisticated interpretation of the data, but this may also limit its use to those cases where the data are sufficiently abundant and exact. For published experiments on oxygen binding to hemoglobin they directly compared the concerted model to different versions of the sequential model. In these direct comparisons the sequential model provides a better fit, but the difference is generally not great, and Koshland et al. stated that in many situations where curve fitting was done it may still not be possible to ascertain the correct mechanism for the enzyme to distinguish the KNF method from the MWC method.[24]

5.4 Conformational Change or Induced Fit?

The MWC model is based on the assumption of conformational change between two states, and there is considerable evidence that proteins may generally have two or more conformations accessible to them, as described in Chap. 2. However, Koshland had perceived a need for more autonomous motions in each subunit in the sequential model for allostery. This KNF model was consistent with his earlier proposal for the need of *induced fit* to explain how enzymes adjust their catalytic sites in binding substrates,[146] by which is meant any movement in the protein after the binding of the first substrate (or first effector ligand) that facilitates the binding of the second substrate, or enhances the access to the substrate of a catalytic residue. Thereafter, these early models have sometimes been interpreted by writers as posing a conceptual contest between concerted conformational change vs. sequential conformational change, or of the process of conformational change vs. the process of induced fit.

The MWC model is too limiting in its requirement that all subunits act in concert, since clear exceptions to this have been observed.[147] But, we also have enzymes where the concerted model provides the best interpretation.[147] This is not surprising, since basic thermodynamic criteria do not rule out either the concerted or the sequential mechanism for allosteric change. For some proteins the concerted model completely satisfies all the observed data, while for others we have evidence that not all subunits have the same conformation, in support of a sequential model. Both models have explanatory power, and we can use them as the experimental data are in accord with their interpretation.

It has been observed that both the MWC model and the KNF model represent limiting subsets of the complete description for possible conformational states, as exemplified by the scheme for a square tetramer in Fig. 5.12. This diagram shows that there are 25 distinct states for this tetramer, whose subunits may adopt two distinct conformations, and these may be empty or binding one substrate, S. For this complete set of all states of the tetrameric enzyme with one type of ligand, the MWC model is limited to the two columns at either end, since only these columns have all four subunits in the same conformation. The standard KNF model is then represented by the set of tetramers along the diagonal of this diagram, which traditionally begins with all subunits in the T conformation. Koshland et al.[24] were aware of the larger set of possible tetramers with mixed conformations and binding, but chose to focus on the simpler model that they hoped would prove to be usable.

Induced fit is not a process that opposes or replaces conformational change. It is simply a separate process that is thought to occur on most or all enzymes. An important distinction for induced fit is that the movement is normally local. Most commonly it involves the movement of one or a few residues in a loop. For the whole protein, movement is not evident or is very modest in extent. This was measured by Gutteridge and Thornton for a set of 60 proteins for which crystal structures exist for an *apo* form, and for a ligand bound form.[148] The result was that for the protein as a whole, observed differences in the backbone structure between the *apo* enzyme and the ligand-bound enzyme are about 0.7 Å on average. However, these authors also compared the crystal structures for all those proteins where the same *apo* form of a specific enzyme had been solved more than once, by the same laboratory or by different laboratories. For 29 such comparisons of an identical enzyme structure, the global differences in the backbone positions of the structure averaged 0.5 Å. This average value of 0.5 Å then represents the noise or uncertainty in such crystal structure determinations. The average value of 0.7 Å for the induced fit comparisons is a value that is just barely above this background, and therefore signifies that modest to no conformational change occurs in the overall global structure during induced fit. The authors concluded that much of the induced fit motion is in the side chains of catalytic residues.[148] Therefore, induced fit represents a different process from the large scale movements between domains that normally account for conformational changes.

We can conclude that induced fit occurs to some limited extent with most enzymes, and while such slight changes in the subunit structure are frequently referred to as different conformations, to avoid confusion with our broader discussion of allosteric enzymes we will treat such movements as part of any enzyme's normal dynamic flexibility. Larger scale conformational changes in regulatory enzymes may be concerted or sequential. We have many examples of each. For enzymes with concerted changes the distinction between the two models disappears, as either one can satisfactorily calculate the V_{max} and K_ms. For those enzymes that do have a sequential order for changing conformation, the KNF model will normally provide better values for these constants.

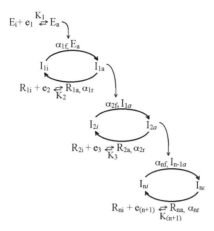

Fig. 5.13. Scheme for an *n*-cycle cascade system. E_i is the inactive forward converter enzyme, which becomes activated via a conformational change to E_a, on binding the effector e_1. The active converter enzyme then phosphorylates the inactive enzyme I_{1i} to form I_{1a}, which continues to the final phosphorylation that activates I_{ni}. Each converter enzyme is opposed by an enzyme having the ability to undue the specific covalent modification, $R_{1i} \ldots R_{ni}$, which also alternate between active and inactive forms that are separately stabilized by additional effectors. [Adapted from Chock and Stadtman, Proc. Natl. Acad. Sci. USA **74**, 2766–2770 (1977)]

5.5 Allosteric Sensitivity and Molecular Switches

The description above for the parameters of positive cooperativity have focused on single enzymes, with the aim of establishing the parameters that are relevant for this type of control, normally by one ligand acting on one enzyme, or protein receptor, or RNA riboswitch. Cells and organisms have many processes that depend on an array of activities to achieve a specific result, and in such more complex processes cooperativity can become even more extended.

Hemoglobin was the first protein whose binding was analyzed by the Hill equation,[123] and a Hill coefficient near 3.2 is frequently observed for this tetrameric protein. Since the hemoglobin tetramer has one of the higher n_H values measured, this leads to the interpretation that the value of n_H approximates the number of subunits in any allosteric oligomer, and that its upper value is limited by this number. Due to the assumptions built into the MWC and KNF models, the equations of these models also limit n_H to be $\leq n$ (the number of subunits in the oligomer). If a sequence of enzymes is coordinated for an end result, then a normal degree of cooperativity by the first enzyme in such a sequence or cascade (Fig. 5.13) is increased by the cooperativity of enzymes that continues this function, leading to much higher cooperativity.

5.5.1 Enzyme Cascades and the Extent of Positive Cooperativity In Vitro

As shown for the enzyme sequence that leads to the release of glucose from glycogen (Fig. 4.2), an enhanced response to an initial stimulus or signal is often achieved by a sequence of enzymes, where each in turn activates the next one, so that an amplified response is possible by the final target enzyme that performs the desired reaction, which is produced by glycogen phosphorylase in Fig. 4.2. To quantitate the kinetics for such an amplified response, Earl Stadtman and colleagues developed a mathematical model to interpret the complexity with the number of enzyme cycles in such a cascade.[149, 150] An example of an n-cycle cascade is shown in Fig. 5.13. Enzyme I_{na} catalyzes the desired reaction, but is initially inactive (I_{ni}). We see enzymes involved in activating I_{na}, with E_i responding to the initial effector molecule (e_1) that acts as the starting signal, and then I_{1a} being formed by the action of E_a, and so forth.

In this example the activating enzymes could act as kinases, since that form of covalent activation is most common. For such an example the model also has several phosphatases to counter the kinases, shown as $R_1,...,R_n$, which also are interconvertible between active and inactive conformations, due to their binding the specific effectors, $e_2,...,e_{(n+1)}$. While up to $n+1$ effector molecules influence the overall rate of I_{na}, the major focus of this analysis will be on the sensitivity of this system to the initial signal, e_1, as it changes in concentration to begin activating the cascade.

The principal outcome of the cascade is the change in activity of enzyme I_n, measured as the proportion of I_n in the form of I_{na}, given as $[I_{na}]/[I_n]$. This involves a binding and catalytic step for each enzyme modification as shown:

$$I_{1i} + E_a \xrightleftharpoons{K_{1f}} E_a I_{1i} \xrightarrow{k_{1f}} E_a + I_{1a}, \qquad (5.29)$$

where K_{1f} is the dissociation constant for binding the effector to the converter enzyme and k_{1f} is the rate constant for the conformational change. The authors also introduced a proportional term for these constants:

$$\alpha_{1f} = \frac{k_{1f}}{K_{1f}}. \qquad (5.30)$$

The expression for the formation of $[I_{2a}]/[I_2]$ in a two-cycle system as a function of the converter enzymes and the concentration of effectors is given by:

$$\frac{[I_{2a}]}{[I_2]} = \left[\frac{\alpha_{1r}\alpha_{2r}(K_1 + [e_1])[R_1][R_2][e_2][e_3]}{\alpha_{1f}\alpha_{2f}(K_2 + [e_2])(K_3 + [e_3])[E][I_1][e_1]} + \frac{\alpha_{2r}[R_2][e_3]}{\alpha_{2f}[I_1](K_3 + [e_3])} + 1 \right]^{-1}. \qquad (5.31)$$

This expands to the full expression for an n-cycle cascade:

$$\frac{[I_{na}]}{[I_n]} = \left[\frac{a_{1r}a_{2r} \cdots a_{nr}(K_1 + [e_1])[R_1][R_2] \cdots [R_n][e_2][e_3] \cdots [e_{n+1}]}{a_{1f}a_{2f} \cdots a_{nf}[E][I_1][I_2] \cdots [I_{(n-1)}][e_1](K_2 + [e_2])(K_3 + [e_3]) \cdots (K_{(n+1)} + [e_{(n+1)}])} \right.$$

$$+ \frac{a_{2r}a_{3r} \cdots a_{nr}[R_2][R_3] \cdots [R_n][e_3][e_4] \cdots [e_{(n+1)}]}{a_{2f}a_{3f} \cdots a_{nf}[I_1][I_2] \cdots [I_{(n-1)}](K_3 + [e_3])(K_4 + [e_4]) \cdots (K_{(n+1)} + [e_{(n+1)}])}$$

$$\left. + \cdots + \frac{a_{nr}[R_n][e_{(n+1)}]}{a_{nf}[I_{(n-1)}](K_{(n+1)} + [e_{(n+1)}])} + 1 \right]^{-1} \qquad (5.32)$$

This now represents the expression for all parameters that influence the activity of enzyme I_{na}. To facilitate calculations for simulating this model, all parameters are initially set equal to 1, and only the concentration of e_1 is varied.

Based on the scheme in Fig. 5.13 and the definitions above, the authors then devised an alternate simplification to express the activity of the key enzyme, I_{na}:

$$\frac{[I_{na}]}{[I_n]} = \left[\left(\frac{K_1}{[e_1]} + 1 \right) \left(\frac{k_r'}{k_f'} \right)^n + \left(\frac{k_r'}{k_f'} \right)^{n-1} + \cdots + \left(\frac{k_r'}{k_f'} \right) + 1 \right]^{-1} \qquad (5.33)$$

Using either (5.32) or (5.33) one can then model the responsiveness of cascades, with an increasing number of cycles, to the variation in the concentration of the initial effector, e_1. These are shown in Fig. 5.14 for cascades including one to four cycles. The important feature of Fig. 5.14 is that the equations for this model make no assumptions about any of the converter enzymes being cooperative in responding to e_1. This is directly evident by examining the change in concentration of e_1 required for I_n to go from 10 to 90% of activation.

However, by adding additional cycles to the cascade, the response pattern shifts to lower concentrations of e_1. The simulation for the published model was designed on a simple numerical scale, without concentration units.[149] To make Fig. 5.14 more realistic, we will examine a simulation with the concentration units on the abscissa at molar. Then, for the one cycle cascade, to change the response from 10 to 90% of maximum requires a change in the concentration of e_1 that goes from about 8 mM to about 600 mM. This is a concentration range that is physiologically very unlikely. The affinity of the responding converter enzymes is at about 100 mM. However, by increasing the system to a four-cycle cascade, the response is much more sensitive, since the same degree of activation now occurs over a range of e_1 from about 8 μM to about 600 μM, with an affinity of about 100 μM by the responding enzymes for e_1. This permits a 1,000-fold increase in sensitivity, and this concentration range for the effector signal is quite normal. Therefore, by having a cascade with several cycles, it is easy for it to be responsive to changes in effectors at their normal physiological concentrations.

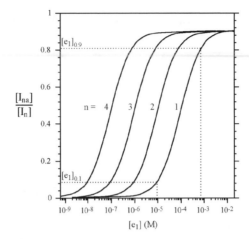

Fig. 5.14. Simulated response curves for the activity of enzyme I_{na}, as a function of the initial effector concentration, and of the number of cycles in the enzyme cascade. Except for $[e_1]$, all other parameters were set equal to 1. The *horizontal dotted lines* show the concentration of the effector required to achieve a 90% change or a 10% change in the active form of I_n. For the case with one enzyme cycle, the *vertical dotted lines* show that the change in $[e_1]$ between 10% and 90% is 81-fold, so that the response of the enzyme in this cascade does not itself have any cooperativity for e_1

Although no individual enzyme needs to show cooperativity for the initial effector, e_1, Stadtman and Chock recognized that the enhanced sensitivity described above was comparable to positive cooperativity.[151] They defined the sensitivity of the cascade by the following ratio:

$$\text{Sensivity} = \frac{8.89[e_{0.5\,max}]}{[e_{0.9\,max}] - [e_{0.1\,max}]} \qquad (5.34)$$

using the concentrations of effector to achieve different proportions of the maximum response. The sensitivity is depicted in Fig. 5.15, with the corresponding n_H values. Figure 5.15 is similar to Fig. 5.2, though the sensitivity plotted on the abscissa of Fig. 5.15, as defined by (5.34) is somewhat different from the sensitivity term used in Fig. 5.2.

An interesting feature of the model by Stadtman and colleagues is that it has no term for the oligomer size of enzymes. Instead of the number of subunits in an allosteric enzyme, the limiting term for the observed cooperativity is the complexity of the cascade, in terms of the number of reversible cycles that form the cascade. The apparent Hill coefficient that may be described is an integral number, and for increased sensitivity this is twice the value of the cycle number (Fig. 5.15). For decreased sensitivity, this relationship is no longer linear. Therefore, for the Stadtman model, n_H values of $2n$ are always possible, and the upper limit for n_H is directly determined by n, the number of cycles in any cascade (Figs. 5.14 and 5.15).

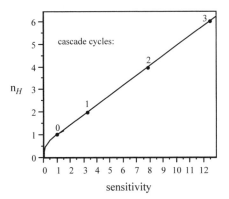

Fig. 5.15. The apparent cooperativity of cascades, with increasing sensitivity provided by a greater number of enzyme cycles. Sensitivity is defined by (5.34)

5.5.2 Enzyme Cascades and the Extent of Positive Cooperativity In Vivo

A well-defined example of how sensitive a biological response may be to a changing stimulus is provided by studies on the ability of oocytes (from *Xenopus laevis*) to initiate maturation in response to a modest change in the concentration of progesterone, the hormone that stimulates this developmental change. To begin the maturation process, oocytes must enter the cell cycle, which requires that the hormone stimulate the mitogen-activated protein kinase (MAPK) cascade.[‡] Therefore, the specific biochemical mechanism is phosphorylation of a target protein.[152] The sensitivity of this process to the available concentration of the hormone is shown in Fig. 5.16. The authors separately measured the in vitro phosphorylation rate of the enzyme, and found it to have a Hill coefficient of 5. But, when they measured the actual response of individual oocytes to this stimulus, the cooperativity to describe this curve yielded an n_H of 42.

While the extent of cooperativity for the protein kinase in vitro is already higher than normal, the remarkable cooperativity for the response in vivo has the feature of providing an on/off switch, so that a predefined, but modest concentration of progesterone achieves an almost universal response from the oocytes. This is attributed to the remarkable cooperativity of the response, since the cascade goes from near 0 to 0.95 V_{max} with an $S_{0.9}/S_{0.1} = 1.11$. This means that at a critical concentration of progesterone where the protein cascade has not yet responded, an increase of only 11% in the concentration of progesterone leads to initiation of the cell cycle in almost all oocytes. The high cooperativity is interpreted as being the cumulative result of the separate protein kinase enzymes plus the new protein synthesis that is involved in oocyte maturation. When inhibitors of protein synthesis were added to the oocytes, then the in vivo response for the MAPK enzymes alone declined to an n_H of 3.

[‡]MAPK enzymes will be described in more detail in Chap. 13.

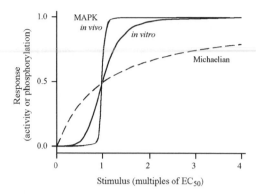

Fig. 5.16. Stimulus response curves for p42 MAPK phosphorylation in vivo ($n_H = 42$) or in vitro ($n_H = 5$). For comparison, a normal curve with no cooperativity is shown. From Ferrel and Machleder, Science, **280**, 895–898 (1998). Reprinted with permission from AAAS

The on/off switch response for the oocytes, mediated by protein kinases, is then comparable to the molecular switches discussed in Chap. 2. Here, the switch is not based on the slowness of the reaction, but on the cumulative effect of several allosteric processes acting in concert.

SECTION 2

K-TYPE ENZYMES

6

HEMOGLOBIN

Summary

The major function of hemoglobin in vertebrates is to bind oxygen in the lungs and transport it to the tissues. Since there is only a threefold to fivefold difference in the normal oxygen concentration between the lungs and the tissues, cooperativity in the affinity for oxygen is necessary to achieve efficient delivery of oxygen. This cooperativity results mainly from a conformational change in the quaternary structure of the hemoglobin tetramer, and is strongly influenced by various effectors. There are two types of effectors. Bisphosphoglycerate (BPG) and inositol hexa(kis)phosphate (IHP) are normally at fairly constant concentrations, and have become agents that weaken the affinity of hemoglobin for oxygen. There are also some appropriate physiological effectors, carbon dioxide and protons, that increase during an anaerobic/anoxic state, and by further lowering the affinity for oxygen, increase the ability of hemoglobin to release oxygen to active tissues that have an increased need for oxidative metabolism.

6.1 Overview of Hemoglobins

6.1.1 The Late Appearance of Oxygen in the Evolution of Life

The earth was formed about 4.4 billion years ago (4.4 Ga), and bacterial life emerged by about 4 Ga. The original planetary atmosphere had no significant oxygen, and life evolved in a reducing environment. This feature remains in current mammalian cells, which internally have a reducing environment, with various cofactors, such as glutathione, to assist in maintaining the proper redox state. The earliest production of oxygen was at about 2.2 Ga,[153, 154] with the appearance of cyanobacteria. The spread of algae and simple plants provided oxygen from photosynthesis, but for the next billion years oxygen remained at low levels, since it was rapidly consumed in the formation of carbon dioxide and its derivatives.

The appearance of oxygen had an immediate impact. First came the requirement for defense mechanisms against reactive oxygen species (superoxide, hydroxide radicals).[*] This was followed in the great majority of organisms by an adaptation of metabolism to benefit from the availability of oxygen as an electron sink. An analysis of 70 genomes shows that for those that became aerobes and facultative anaerobes, at least 1,000 new chemical reactions dependent on oxygen became possible, as judged by the new genes that evolved, in comparison to anaerobes. For most microbes their genomes expanded 10–20% with these new genes.[154] For mammals, more than one-third of our genes code for proteins related to oxygen metabolism.[154]

Fungi and algae were abundant by 1 Ga, and they, along with mosses and simple plants covered suitable land areas. Although total photosynthesis had increased, the additional oxygen was just as quickly consumed. However, the combination of cellular growth on and in surface soils, combined with long-term weathering effects resulted in increased formation of clays. These clays have been shown to bind and sequester carbon compounds.[153] Therefore, a steady increase in such clay layers by 0.8–0.6 Ga led to a dramatic increase in atmospheric oxygen. And this time coincides with the beginning of the metazoan radiation of larger complex organisms. Larger organisms cannot function without adequate oxygen, and therefore evolved only when the available atmospheric oxygen made this possible. In less than a billion years the earth's atmosphere changed from containing no significant oxygen, to containing approximately 21% oxygen today.

6.1.2 The Diversity of Globins

Hemoglobins are a subset of the larger globin family, with small subunits that most commonly have seven or eight alpha helices and fold so as to form a binding pocket for a porphyrin ring, or heme (Fig. 6.1). Heme is a porphyrin plus covalently bound iron, which serves as the cofactor for binding oxygen. Globins are found in all species examined from all kingdoms of life. Globins appear to be ancient, and existed before oxygen became an abundant atmospheric gas less than 1 billion years ago.

Since the early earth had an anoxic, reducing environment, the earliest life forms evolved in the absence of oxygen. As oxygen began to increase in the environment, it would initially have been toxic, and the danger of oxygen radicals was described in Chap. 2.1.2. This supports the assumption that the earliest oxygen binding proteins may have served a protective function, by sequestering oxygen, similar to the function of leghemoglobin today. This initial function for hemoglobin is consistent with the fact that proteins similar to hemoglobin occurred so early, and apparently in all life forms.

However, oxygen is also very useful as an electron acceptor that greatly enhances the energy available from carbohydrates, and the adaptation of existing globins to oxygen binding and oxygen transport was an evolutionary success. Globins also have a good affinity for binding nitric oxide (NO), and it is possible that some of the earliest globins evolved to bind and sequester this molecule, since initially it would have been toxic. Through adaptive changes NO came to have a signaling function, and the release of NO

---

[*] A brief description of two enzymes involved in the removal of reactive oxygen species occurs in Chap. 2.1.2, where we focused on the slowest enzymatic reaction rates that would still be compatible with the survival of simple microbes.

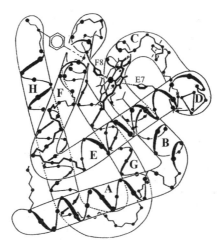

Fig. 6.1. A model of the globin fold. By convention the helices of such globins are defined, and their residues numbered, as for the first myoglobin structure. The heme ring is covalently bound via an invariant histidine (F8) and a second residue at E7 that is frequently a histidine

by Hb in the smaller blood vessels has been shown to alter blood vessel constriction, and thereby improve oxygen delivery (described in Chap. 6.4).

The currently observed diversity of hemoproteins is consistent with these proteins existing very early in evolution. Porphyrins may have been among the earliest molecules, since their synthesis is not complicated. Porphyrins readily bind certain divalent metals. With magnesium they form chlorophyll; with iron they form cytochrome or heme. The need for electron carriers suggests that cytochromes may have been among the earliest hemoproteins, and that the evolution of chlorophyll and photosynthetic proteins followed, and finally hemoglobins became necessary with the early oxygen atmosphere.

6.1.3 Diversity of Globin Structures and Kinetics

Based on the comparison of the amino acid sequences for known globins, it is very clear that they are all descended from a common ancestor.[155, 156] It is also evident that this evolutionary divergence was accompanied by various modifications for many individual globins, to make them more effective in their specific organisms. This is briefly shown in Table 6.1, where only a few hemoglobins are listed, with the purpose of showing the variety in oligomeric assembly, as well as the range of affinity for oxygen. The examples in this table are listed by their evolutionary time for appearance, with the bacterial globin being the oldest and the human globins the newest.

There is a remarkable range for the assembly of globin subunits. In sperm whale, myoglobin is normally monomeric, but hemoglobins form simple dimers in bacteria and clams. Humans have a gene duplication leading to the slightly altered beta globin, and the highly cooperative human hemoglobin in erythrocytes is an $\alpha_2\beta_2$ tetramer. In the parasitic worm *Ascaris*, a gene duplication plus fusion led to the protein having two very similar domains on a single subunit. One such subunit would be structurally comparable to the

Table 6.1. Globin quaternary assemblies and kinetics

Species	Oligomer	p50 (torr)	n_{Hmax}
Bacteria, *Vitreoscilla* hemoglobin[a]	α_2	0.02	
Parasitic worm, *Ascaris suum* hemoglobin[b]	$(\alpha-\alpha)_8$	0.004	1
Earthworm, *Lumbricus terrestris* erythrocruorin[b]	$(\alpha_{12})_{12}(\beta_3)_{12}$	1.6–17.3	2.5–7.9
Clam, *Scapharca inaequivalvis* hemoglobin[b]	α_2	4.8	1.5
Sperm whale myoglobin[b]	α	0.5	1
Human hemoglobin[b]	$\alpha_2\beta_2$	1.6–18.8	2.5–3

[a]From Giangiacomo et al.[157]
[b]From WE Royer, Jr.[158]

dimers found in other organisms. These in turn form the active octamer, which is then comparable to a complex of four human hemoglobin tetramers. Earthworms have a variant globin, known as erythrocruorin, and this forms a huge complex. The actual hemoprotein subunits assemble into dodecamers, and 12 dodecamers are stabilized in a much larger complex by the trimeric β linker protein. This linker protein occurs at the corners of the complex, where each trimer binds with three hemoprotein dodecamers, along the three axes of a normal cube. This results in the total assembly containing 180 subunits.

It is also evident that the *Vitreoscilla* and *Ascaris* globins have the greatest affinity for oxygen, consistent with a function for sequestering oxygen. Whale myoglobin has an intermediate affinity, and earthworms, clams and humans have the weakest binding of oxygen, which is more appropriate for proteins that must transport and then release oxygen. While cooperativity is such an important feature of mammalian hemoglobin, we see that this allosteric feature is not inherent in all globins, since several have an n_H of 1. Allosteric regulation was a later adaptation that occurred in some globins, and involved different, but functionally relevant regulatory molecules in different organisms.

6.1.4 Hemoglobin and its History in Understanding Cooperative Binding

It has long been clear that hemoglobin's major function is the transport of oxygen from a source, such as the lungs of mammals, to a separate internal site where it is required, such as our peripheral tissues. This transport function requires only the binding of the principal ligand, oxygen, with no chemical alteration. Therefore hemoglobin is not normally considered as an enzyme, even though we will see examples of enzymatic activity by hemoglobin later. Several properties of hemoglobin have made it a protein that is easy to study:

1. The importance of oxygen transport was understood very early, and population studies showed variations in blood types with related changes in stability and transport efficiency.
2. Blood generally contains erythrocytes at almost 50% of its total volume, and mammalian erythrocytes contain hemoglobin at the high concentration of about 25 Mm,[†] making it very easy to obtain and purify in abundance.

[†] This high concentration must be compared to the average concentration of a normal protein, at about 1 μM.

3. A significant change in the absorbance of oxyhemoglobin, compared to deoxyhemoglobin,[‡] made it possible to follow the oxidation of hemoglobin by spectrophotometric techniques available at the end of the nineteenth century.
4. While animals were the normal source of blood in early studies, it was still possible to obtain adequate human blood samples, since experimenters could use their own blood.

The above facts led to hemoglobin being the molecule with which various biochemical measurements were first accomplished. In 1904 it was the first protein shown, by Christian Bohr and colleagues, to have sigmoidal curves for the binding of a ligand.[159] In 1910 this novel binding curve led to the first mathematical description for such cooperative binding with the Hill equation, and the assignment of the Hill coefficient for the extent of cooperativity.[122] In 1965 it was the model for testing the MWC theory (Monod, Wyman, and Changeux) for concerted cooperativity,[49] and the KNF theory (Koshland, Némethy, and Filmer) for sequential cooperativity.[24] In 1968 it was the first oligomeric protein that produced a crystal structure.[160]

6.1.5 Types of Globins and Their Normal Functions

Table 6.2 summarizes various functions demonstrated for the range of known globin molecules that have been studied. Since oxygen is a gas, the affinity is defined by the relative pressure that produces 50% binding (p50) of oxygen by the hemoglobins. The main function of leghemoglobin, so called because it is found in legumes (plants), is to sequester oxygen in the plant root nodules where conversion of nitrogen gas to nitrate (also called nitrogen fixation) would be inhibited by molecular oxygen. This function explains why it has the lowest p50 (= tightest binding) for oxygen. Flavoglobins are found mainly in microbes. They have a flavin adenine dinucleotide (FAD) binding domain as well as a normal hemoglobin-like domain. The toxicity of nitric oxide (NO) is prevented when the binding of oxygen activates the FAD domain for NO reduction via FAD.[161]

Hemoglobin (Hb) from higher vertebrates such as the human adult HbA, have two functions: the well established function of transporting oxygen, and an additional function of using nitric oxide (NO) as a signaling molecule to influence blood vessel constriction (see Chap. 6.4). Until recently myoglobin was normally considered to be an oxygen storage protein. This interpretation was due to the high concentration of this protein in muscle (Table 6.2) and its high affinity for oxygen. These facts were combined to support a model of myoglobin holding on to oxygen tightly until the muscle's demand for oxygen, due to increased activity, caused muscle oxygen concentrations to fall enough that the release of oxygen by the tight binding myoglobin permitted an extra amount of effort. Such an interpretation was sometimes emphasized in the context of the "fight-or-flight" scenario, with the view that this stored oxygen permitted an animal to survive due to the extra effort made possible.

While one may still see the large amount of myoglobin in muscle as providing a reserve of oxygen, it is now established that myoglobin also has a transport function, picking up oxygen at the membrane of a muscle cell, and then conveying it within the

---

[‡] This is evident by arterial blood being red, while veins are more blue.

Table 6.2. Functions of hemoglobin and other globins

Type (source)	Function	p50 (torr)[a]	Concentration (μM)	Refs.
Leghemoglobin (root nodules)	O_2 scavenger	0.03		162
Flavoglobin (S. cerevisiae)	NO removal	0.02		163
Hemoglobin HbA (erythrocytes)	1. O_2 transport to tissues 2. NO transport/sign aling	9.8; 26[b]	25,000	164, 165
Myoglobin (muscle)	O_2 transport to mitochondria	~0.9	400	166
Cytoglobin (peripheral tissues)	O_2 transport to mitochondria	5.4	~1	167
Neuroglobin (peripheral tissues)	O_2 transport to mitochondria	7.5	1	168, 169

[a] Affinities for oxygen of stripped hemoglobin
[b] p50 is increased by heterotropic effectors

cell to a mitochondrion, where the oxygen is consumed in generating ATP. The higher affinity for oxygen would guarantee that as oxygen, released by passing erythrocytes, diffuses across a cell membrane it will be bound and retained within the cell. Tissue concentrations of oxygen are generally in the range 19–37 torr.[170] This would be an average oxygen concentration of about 36 μM. If muscle myosin occurs at a concentration of 400 μM (Table 6.2), then under steady state conditions perhaps one-half of the myoglobin contains oxygen, and the remaining molecules are empty, having released their oxygen at the mitochondrial membrane where the immediate oxygen concentration is below micromolar. This concentration is similar to the affinity constant for myoglobin, so that release of ≥ 50% of bound oxygen occurs readily at the mitochondrial surface.

Therefore, myoglobin is also an oxygen transporter, similar to hemoglobin. But, unlike hemoglobin, it functions within a cell where the general concentrations of free oxygen at the cell membrane is always 50–100-fold greater than at the mitochondrion. Therefore, the 2-log rule for binding assures that the normal affinity of myoglobin leads to almost complete loading at the cell membrane, followed by at least 50% unloading at the mitochondrion. This makes myoglobin an effective transporter since it operates between two regions of appropriately different oxygen concentrations.

Cytoglobin and neuroglobin are not as well characterized. Cytoglobin is found in most tissues, and may have a function similar to myoglobin. It occurs at a much lower concentrations in cells, but most cells do not have as significant a requirement for oxygen as muscle. Both cytoglobin and neuroglobin have an oxygen affinity that is somewhat lower than myoglobin's, but still strong enough to function as an intracellular transporter. Although present at lower concentrations, cytoglobin is cooperative in binding oxygen, and therefore would be a somewhat more efficient intracellular transporter. A reason for having cytoglobin as a transporter in most tissues, and myoglobin only in muscle, is that this evolution of the myoglobin gene would permit it to be uniquely regulated in muscle cells, assuring a much higher concentration of the required oxygen carrier in these cells. This is consistent with the evolution of other isozymes mentioned in Chap. 1.

Table 6.3. Molar binding constants for human globins at $37°C^a$

Globin	Oligomer	Affinity for		
		O_2 (torr)	O_2 (μM)	CO (μM)[b]
Hemoglobin A, HbA	$\alpha_2\beta_2$	9.8^c	13.2	0.13
Hemoglobin F, HbF	$\alpha_2\gamma_2$	7^c	9.5	0.09
Myoglobin, Mb	α	0.9^d	1.2	
Neuroglobin, Ngb	α	7.5^e	10.1	
Cytoglobin, Cygb	α_2	5.4^e	7.3	

[a] For hemoglobin A and F, O_2 affinities are for stripped hemoglobin
[b] Di Cera et al.[167]
[c] Chen et al.[172]
[d] Brunori et al.[166]
[e] Fago et al.[173]

Neuroglobin was named for being most abundant in neural cells, and is chiefly found in the brain, though it has very varied cell concentrations in different brain regions. Because cytoglobin is also in these tissues, a role for neuroglobin may be to function during hypoxia. Studies with rats subjected to hypoxia showed that survival was improved by overexpression of neuroglobin, and the functional outcome was worse if the rats were given antisense RNA to lower the expression of neuroglobin.[174] Oxygen free radicals and peroxynitrite become more abundant with oxidative stress during anoxia. The increasing neuroglobin concentrations that accompany this state could then serve as detoxifying agents by sequestering these reactive species.[175]

Because neuroglobin and cytoglobin were only recently discovered, one may also perceive them as being more recently evolved. Enough sequences have now been obtained that alignments of these sequences clearly support the conclusion that cytoglobin and neuroglobin are older.[176] Their amino acid sequences have changed less over evolutionary time, suggesting the conservation of crucial, and possibly ancient physiological functions.

6.1.6 Converting Binding Constants to Molar Units and for 37°C and pH 7.4

It is very traditional to represent the concentration of oxygen in units of mm Hg, or torr, since oxygen is a gas. To be consistent with the literature on hemoglobin, all affinity values will initially be presented in these units. To really appreciate the strength of such binding interactions, and to compare them to the enzymes in other chapters, it is more useful to convert oxygen pressure to units of molarity. Many experimental determinations were done at "room temperature" which varies from 15 to 30°C, depending on the actual study. These have been recalculated for 37°C. Also, the pH employed has varied from 6.5 to 7.5, and such values have been recalculated to pH 7.4. Examples of how the p50 varies with these experimental alterations will be described in Chap. 6.2.2.

For the principal globins, such recalculated values are shown in Table 6.3. Also, a molarity scale is at the top of figures for oxygen binding curves. One can therefore easily see the correspondence between the units of torr (mmHg), and the more standard units of molarity used in enzymology. It is then evident that the affinity of globins for oxygen is normally in the low micromolar range, a value that shows strong binding, and is very comparable to many metabolic enzymes. However, the *Ascaris* hemoglobin (Table 6.1)

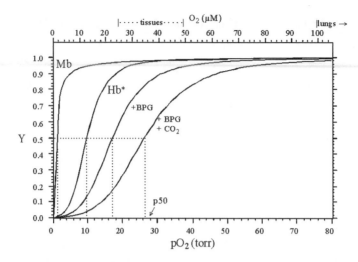

Fig. 6.2. Oxygen binding curves for human hemoglobin (Hb*, stripped hemoglobin) and myoglobin (Mb). The two *curves* at right are for stripped hemoglobin in the presence of 2,3-bisphosphoglycerate (BPG) and with CO_2 added. Hemoglobin curves were produced using the kinetic values of Imai,[177] and recalculated for 37°C

then has a p50 of 4 nM, and we see that this very strong binding constant is completely consistent with a role for sequestering oxygen in a cell that is required to be anaerobic.

We also have binding values for carbon monoxide (Table 6.3). For both human adult and fetal hemoglobin, carbon monoxide binds about 100-fold more tightly. This results in a very slow off rate and explains the toxicity of this gas.

6.2 Cooperativity

To understand both the need for, and the benefit of, cooperativity, let us consider an erythrocyte during normal circulation. We need a carrier that has the appropriate affinity to become fully oxygenated in the alveoli of the lungs, and then largely deoxygenated (unloaded) in the capillaries of the tissues. At sea level the ambient air has a partial pressure (pO_2) of 159 torr. Respiring animals cannot completely empty the lungs when exhaling, and the inhaled air is thus mixed with the remaining air, and therefore alveolar air has a pO_2 of about 100–120 torr, and this is the oxygen concentration at which hemoglobin must bind oxygen. Cells in our tissues tend to have pO_2 values that are generally ≤40 torr, though working muscle can fall below 20 torr (Fig. 6.2).

For such binding curves, Y represents the fraction of hemoglobin with bound oxygen:

$$Y = \frac{Hb \cdot O_2}{Hb_{total}}. \tag{6.1}$$

Using the 2-log rule, we need a transporter with a p50 (the affinity of hemoglobin for oxygen) of about 10 torr to load up in the lungs, and with a p50 of about 300 torr to fully unload in the tissues. Clearly, any protein with a constant affinity will not satisfy such concentration limits, and would be a poor delivery vehicle. We then see that myoglobin would not be suitable for a transport function in the blood. While its p50 value assures full loading of oxygen in the lungs, it would not release significant oxygen in the peripheral tissues, where the local pO_2 is often near 40 torr. If we make the affinity appropriately weaker, to assure unloading in the tissues (i.e., an affinity of at least 100 torr, or even weaker), then this molecule will not be able to bind as much oxygen in the lungs.

The remarkable solution to this important problem is to have an allosteric transporter that has an appropriately high affinity (10 torr, or lower) in the lungs, and an appropriately low affinity (300 torr, or higher) in the deoxygenated capillaries, and such a change in affinity is demonstrated by hemoglobin (Fig. 6.2). Having distinctly different affinities in two distinctly different environments is largely due to the presence of 2, 3-bisphosphoglycerate (BPG). As shown in Tables 6.1 and 6.2, hemoglobins and other globins generally have low p50 values (tight binding), which is consistent with their evolution from an ancestor that functioned to sequester toxic oxygen. It is the binding of the regulatory effector BPG that significantly weakens the affinity for oxygen, and thereby favors the release of oxygen. The affinity for stripped hemoglobin represents the affinity of the R conformation, and this R conformation is stabilized by bound oxygen, and is then the conformation of hemoglobin as it circulates through the lungs.

It must be remembered that hemoglobin molecules are never completely empty. Thus, when entering the lungs most hemoglobin molecules normally still carry 2 or 3 oxygens per tetramer, and being in the R conformation, with an affinity near 10 torr, load almost 100% with oxygen before returning to the tissues. The critical function of hemoglobin is to deliver this bound oxygen to the tissues, and to do this it must release the oxygen. Since binding of oxygen represents a rapidly reversible process, the release or the binding of oxygen is directly related to the surrounding oxygen concentration. At the lower pO_2 of the tissues, with the release of one O_2 the conformational change to the T state becomes more favorable. It is the T conformation that more avidly binds BPG, and the binding of this regulatory effector stabilizes the T conformation. Since the T conformation has a much weaker affinity for oxygen (p50 values approach 600 torr) it is then able to release oxygen, since the surrounding pO_2 is near 40 torr.

For a normal mixture of hemoglobin with some molecules in either of the two conformations, the p50 value of 26 torr (Table 6.2) is not a true binding constant. It represents the average affinity for the ensemble of molecules. The fact that this average p50 is not the average of the separate p50 values for the distinct conformers, defined above, reflects the fact that the two conformers are not present at equal concentrations for the conditions at which the binding curve was determined. Hemoglobin has also become able to bind CO_2 in respiring tissues, and this further stabilizes the T conformation, and thus the average affinity is weakened (Fig. 6.2).

To emphasize the benefit of this allosteric mechanism with two very different affinities, Fig. 6.3 presents a simulated globin with a constant affinity identical to the p50 of normal human hemoglobin. We will also assume that this simulated globin is a tetramer, for an easier comparison with hemoglobin. We then see that this globin would not load up fully with oxygen in the lungs, reaching a Y value of about 75% (three subunits loaded with O_2). In the tissues, the average oxygen pressure near 30 torr means that the

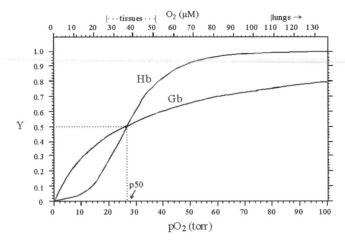

Fig. 6.3. Limited oxygen delivery by a nonallosteric oxygen transporter in blood. Hb represents normal human hemoglobin. Gb represents a simulated oxygen binding curve for a globin with constant affinity for oxygen, equal to the average p50 of hemoglobin

globin would still have almost 50% bound oxygen (two subunits per tetramer). Therefore, such a globin transporter would be able to deliver one O_2 per tetramer for each passage through the circulation. But the binding curve for hemoglobin shows that it should always be able to deliver at least two O_2 per hemoglobin tetramer, and the steeper hemoglobin curve shows that if tissue O_2 concentrations were lower, such as during strenuous work, the hemoglobin transporter might unload possibly a third O_2 of the four being carried.

By comparison, the binding curve for the globin shows that it would unload only a modest amount of additional oxygen under increased demands by the tissues. Although hemoglobin (Hb) clearly looks superior, one may interpret that the simulated globin (Gb) would still be a good transporter. However, the above description was done in the context of binding as an equilibrium process, which is true for normal solutions in a defined container. Within an organism, equilibrium binding is definitely not applicable to hemoglobin, though we can still use it for purposes of comparison. The physiology and mechanics of circulation are such that both binding and release of oxygen should be seen as more dynamic processes, where the kinetics of the on and off steps have more relevance. This becomes more evident with the fact that the oxygen binding curves (Fig. 6.2) clearly imply that at the oxygen concentration of the tissues, hemoglobin should release at least 50% of the bound oxygen. In mammals it has also been observed that erythrocytes returning to the lungs are normally still 75% filled with oxygen. On average, erythrocytes pass through tissue capillaries in ≤1 s, not enough time to approach any equilibrium for oxygen binding and release. In this more dynamic context, the allosteric features make hemoglobin decidedly superior, since it must be able to deliver at least twice as much oxygen per tetramer in this very limited amount of time.

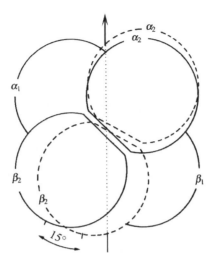

Fig. 6.4. Reversible conformational change from T to R. Perutz calculated that $\Delta G^{T \to R} = 3.5$ kcal/subunit.[178] This is the energy provided by ionic bonds, such as those found to stabilize the T conformation (*solid line*), which become disrupted in the R conformation (*dashed line*). The major conformational change is the rotation of the $\alpha_2\beta_2$ dimer by 15° relative to the $\alpha_1\beta_1$ dimer

6.2.1 Structural Mechanism for Cooperativity

Crystal structures were determined by Perutz and colleagues for both the R and the T conformation of hemoglobin. Comparison of the two provides an understanding of the changes leading from one structural state to the other.[178] One important change involves the position of the iron relative to the heme. In the deoxy state (T), the iron is about 0.5 Å above the plane of the heme. With the binding of oxygen, the iron moves toward the plane of the heme, and pulls on the proximal histidine to which it is bound. This affects the helix containing the histidine, and this movement ruptures ionic interactions between the $\alpha_1\beta_1$ and $\alpha_2\beta_2$ dimers.[§] These are the bonds that stabilize the T conformation. Once broken, there is a rotation of the $\alpha_2\beta_2$ dimer by 15° relative to the $\alpha_1\beta_1$ dimer (Fig. 6.4).

6.2.2 Regulatory Effectors

6.2.2.1 Bohr Effect

The most common effectors are those that define the normal plasma or intracellular environment. Of these, the most likely to be important, due to their ability to vary as a function of nutritional state or health, include pH, carbon dioxide, and temperature. Although osmolar changes are less likely in vivo, the ion concentration also influences the overall conformation of the globins, and therefore their affinity for oxygen.

[§]The term $\alpha_2\beta_2$ normally defines the hemoglobin tetramer, with two alpha and two beta subunits. In discussing the protein structure, $\alpha_2\beta_2$ refers to the second alpha–beta dimer.

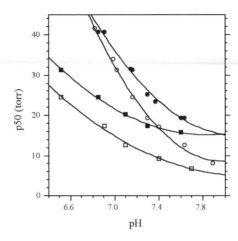

Fig. 6.5. The Bohr effect is the change in affinity due to protonation of hemoglobin. (*Open square*) Stripped hemoglobin; (*filled square*) Hb plus CO_2; (*open circle*) Hb plus saturating 2,3-bis-phosphoglycerate (BPG); (*filled circle*) Hb plus BPG and CO_2. Data from Kilmartin and Rossi-Bernardi[179]

Protonation of hemoglobin has a direct affect on hemoglobin's affinity for oxygen. This was first observed by Christian Bohr and colleagues,[159] and it is still known as the Bohr effect. Their original measurements noted the change in affinity with increasing carbon dioxide gas administered to hemoglobin solutions, but because carbon dioxide is converted to bicarbonate and protons, it was understood that protons could readily bind with some amino acid side chains. It is now known that such protonation occurs principally at the histidine residues, since these have a pK_a in the physiological range.

The Bohr effect is illustrated in Fig. 6.5, where we see that for stripped human hemoglobin the affinity becomes more constant at $pH \geq 7.8$, but begins to steadily become weaker as the solution becomes more acidic. The acidic pH range used in these experiments (Fig. 6.5) may be achieved under physiological conditions in mammals, if limited to some tissues, such as muscle. Since actively functioning muscle will switch to anaerobic glycolysis to maintain the highest rate of work, this transient metabolic state will be accompanied by increasing acidity, as lactic acid formation from pyruvate permits the regeneration of NAD from NADH. Since NAD is the limiting component for this form of glycolysis, mild acidity permits continued muscle function under strenuous conditions.

Normal oxygen binding curves at three pH values are shown in Fig. 6.6. While we tend to focus on the displacement of the normal curve at pH 7.4, either to the left (pH 7.6) or to the right (pH 7.2), it is the change in saturation along the *y*-axis that signifies the difference in oxygen unloading. Therefore, at any given pO_2 for the tissues, hemoglobin

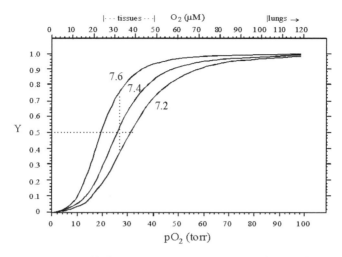

Fig. 6.6. Oxygen binding at 37° by human hemoglobin as a function of pH

will unload about 10% less at the higher pH, and about 10% more at the lower pH. This is exactly what the tissues require, since acidity is a signal that ATP consumption is exceeding oxygen availability.

6.2.2.2 Temperature

Temperature influences the dynamic flexibility of all proteins, and therefore the binding of their normal ligands. We see a weakening of oxygen binding as the temperature is raised for both human neuroglobin and cytoglobin (Fig. 6.7) with a change of about twofold in affinity between 25° and 37°. One must be aware of this change in affinity,since almost all studies of oxygen binding are done at room temperatures below 37°. The observed change in affinity for oxygen (Fig. 6.7) is clearly nonlinear, and it is therefore quite standard to display such temperature data as an Arrhenius plot, as shown for human hemoglobin (Fig. 6.8).

The Arrhenius plot makes linear the observed relation between p50 and temperature. The data in Fig. 6.8 were obtained at lower temperatures, and can be extrapolated to 37° for a very good estimate of the p50 at this normal body temperature. The extent to which such extrapolations may be made is limited, since much higher temperatures often change a rate limiting step in a catalytic reaction. But, for a simple binding interaction, over such a modest increase in temperature, this extrapolation is generally valid, because the Arrhenius plot shows binding to be linear.

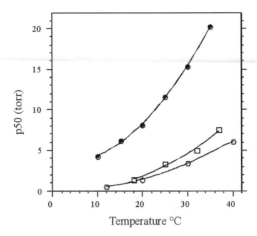

Fig. 6.7. Change in oxygen affinity at pH 7.4 with temperature, for human neuroglobin (*open square*), cytoglobin (*open circle*), (data from Fago et al.[173]) and hemoglobin (*filled circle*). Data from Imai and Yonetani[180]

The experiment that produced the data for Fig. 6.8 was done with individual crystals of hemoglobin in the deoxy (*T*) state.[181] This special experimental format had the desired benefit of guaranteeing that all the molecules would be in the same conformation, which is never achieved in a solution of hemoglobin. It also demonstrates, that while oxygen binding normally induces the conformational transition from *T* to *R*, that did not occur for this experiment, since the crystals did not crack. Also, the observed binding values showed no evidence of cooperativity, because the Hill coefficient was 0.99.

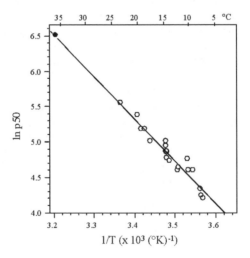

Fig. 6.8. Change in oxygen affinity of human hemoglobin with temperature. (*Open circle*) Data from Rivetti et al.[181] for oxygen binding to a single crystal of deoxy-hemoglobin; (*filled circle*) extrapolated value for 37°C. The p50 at 37° = 650 torr, which is an estimated limit for the weak affinity of the T conformation

6.2.3 Special Regulatory Effectors

The effective transport of oxygen to the tissues is very important for the energy requirements of active animals. To optimize this process, humans have a circulatory system with a volume of 4–5 L. Erythrocytes account for 39–50% of this volume, where 50% is near the upper limit possible to maintain normal blood pressure. Erythrocytes themselves have become simple bags of hemoglobin. They are unusual among cells in having neither a nucleus nor any mitochondria, as these organelles require too much space within the cell. Within the cells hemoglobin is present at a concentration of about 25 mM, the largest such value for a single protein.

Since the ancestral globins had affinities for oxygen that were too strong to be ideal transporters, additional effector molecules have become adapted to binding hemoglobin and weakening its affinity for oxygen. The most significant are 2,3-bisphosphoglycerate (BPG) which is the principal effector in animals, and IHP, with the same role in birds. These two effectors have similar binding sites between the two beta subunits, and therefore are normally found at a stoichiometry of 0.9–1 per hemoglobin tetramer. The lower value is most common, and only organisms living at higher altitudes normally have a stoichiometry greater than 0.9. These effectors bind more readily to the T conformation, as the rotation of one dimer in the tetramer makes a binding site between the beta subunits more accessible. They bind reasonable well to the T conformation with K_d values of 2×10^{-5} M (BPG)[182] and 7×10^{-5} M (IHP)[183].

The effect of BPG is evident in the preceding figures with oxygen binding curves (Figs. 6.2, 6.4, and 6.6). This effector weakens binding by at least a factor of two, with some variation due to additional variables such as pH and temperature. The instability of BPG limits the storage time for human blood to less than 1 week. The normal concentration of BPG in fresh erythrocytes is about 4–5 mM, but it hydrolyzes easily and in a stored vial this concentration approaches zero by 14 days of storage.[182] In the absence of this effector, the stored blood then resumes the affinity observed with stripped hemoglobin (Fig. 6.2),

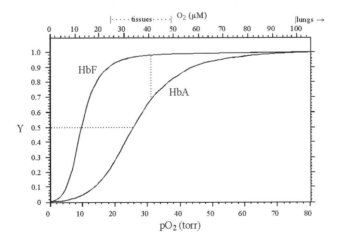

Fig. 6.9. Fetal hemoglobin (HbF) has stronger affinity for oxygen than adult hemoglobin (HbA). At a normal pO_2, about 25% of oxygen on HbA is transferred to HbF

and thereby loses much of its ability to function as a transporter, since it less effectively releases the bound oxygen to the tissues.

6.2.4 Fetal Hemoglobin

Fetal hemoglobin (HbF) contains a different type of subunit, gamma, to replace the beta subunits. These two types of subunits still form a cooperative tetramer. However, this change causes HbF to have a very limited ability to bind BPG, and therefore it maintains the affinity of a stripped hemoglobin. The benefit of this is shown in Fig. 6.9. At the pO_2 found at the placenta, the difference in affinities results in the transfer of about one oxygen per tetramer from the maternal hemoglobin (HbA). Some fraction of the bound oxygen on HbF will then be released to the fetal tissues. HbF is not as efficient a transporter within the fetal circulation, compared to HbA. But, the need to successfully acquire oxygen from HbA requires the stronger affinity.

6.3 Cooperativity Measured by Different Models

The remarkable number of studies with hemoglobin has provided sufficient data to use this molecule as a test for models of cooperativity. As discussed in Chap. 5, cooperative binding may occur in a concerted fashion, which defines a two-state model. It may also be defined as being sequential, with three or four binding states, for the four separate subunits in a hemoglobin tetramer. Using actual oxygen binding data from a publication by Imai[177] , and the equations shown in Table 6.4, simulated curves were derived to test how well they fit the actual data points (Fig. 6.10).

The Adair equation assumes independent binding by each subunit. It therefore represents a sequential model for interaction of subunits in the binding of oxygen, and is comparable to the more general KNF equation described in Chap. 5. It was originally developed as an empirical equation by Adair, specifically for defining oxygen binding to hemoglobin,[184] and is used here because the appropriate constants for the experiment in Fig. 6.10 were available. By iterative calculations these values for the separate kinetic

Table 6.4. Models for simulating cooperative oxygen binding[a]

No. of binding states	Model	Equation	Refs.
Sequential: 4	Adair	$Y = \dfrac{4K_1p + 12K_1K_2p^2 + 12K_1K_2K_3p^3 + 4K_1K_2K_3K_4p^4}{4(1 + 4K_1p + 6K_1K_2p^2 + 4K_1K_2K_3p^3 + K_1K_2K_3K_4p^4)}$	184
Concerted: 2	MWC	$Y = \dfrac{LcK_Rp(1+cK_Rp)^3 + K_Rp(1+K_Rp)^3}{L(1+cK_Rp)^4 + (1+K_Rp)^4}$	23
Concerted: 2	Hill	$Y = \dfrac{p^n}{p50 + p^n}$	123

[a]The original published equations have been modified to use the same terms: Y is the saturation function for hemoglobin with oxygen; p is the concentration of oxygen; K is the affinity constant; and the subscript denotes which subunit (Adair), or conformation (MWC).

constants were determined, and the curve so obtained (Fig. 6.10A) clearly makes an excellent fit to the oxygen binding data.

The MWC equation is designed for a two-state binding system, with all subunits having either a weak binding constant (K_T) or a strong binding constant (K_R). They are expressed as c ((5.1) and Table 6.4) which defines the ratio of these two constants. L defines the ratio of the two conformational species (5.5). While the Adair equation provides an excellent fit (Fig. 6.10A), the simpler MWC equation still provides a very good fit. This can be tested mathematically, but is also evident to the eye (Fig. 6.10B). This fit is considered good enough since the majority of papers on oxygen binding by hemoglobin have employed the MWC equation to define the experimental data.

The simplest simulation for a two-state model is the Hill equation. Whereas the MWC equation includes two affinity constants, for the two different conformational species, the

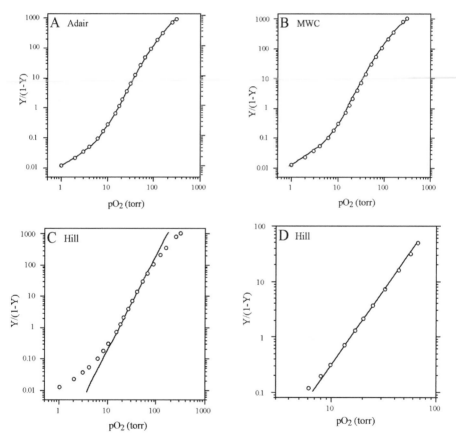

Fig. 6.10. Three simulations to fit a data set for oxygen binding by human hemoglobin, at pH 7.4, and 25°. The continuous curves were made with (**A**) the Adair equation, (**B**) the MWC equation, and (**C, D**) the Hill equation. Data from Imai[177]

Hill equation has simplified this by having only a single term for affinity, the p50 value which represents the average affinity for the ensemble of T and R conformations. It is immediately evident that the fit of this simulation is quite deviant at the lower and upper concentrations of oxygen (Fig. 6.10C). The Hill equation defines a straight line on a log–log plot, and therefore can only help to define the data set for the middle of the ligand concentration range. It is the slope of this line that provides the Hill coefficient, which has a value of 3.09 for the data of Fig. 6.10C.

Note that the oxygen concentration range tested in this data set goes over almost three orders of magnitude. This is an exceptionally wide range, since the majority of enzyme kinetic studies are performed over a range of substrate concentration that is generally far below 100-fold. Hill first developed this equation to define the cooperative binding of hemoglobin for a data set with oxygen values ranging from 10 to 100 torr.[123] For this data set his equation provided very good agreement. When we take only the central range of oxygen values in Fig. 6.10C, we avoid the deviations at the ends of the curve, and now also see that the Hill equation gives a reasonable fit to the data of Fig. 6.10D, for a tenfold change in oxygen concentration. Note the change in scale for this last figure along both axes. By expanding this part of the figure, we again see that the Hill fit is not as good as the Adair fit, even over this more limited oxygen concentration range.

What the overall results of Fig. 6.10 demonstrate is that the best fit to the data points are defined by an equation that has the most terms for affinity, with a different binding constant for each of the four hemoglobin subunits. As our modeling is simplified to using only two affinity values in the MWC equation, the curve fit is still good, but is a little less exact. When we simplify our modeling to a single, average affinity term in the Hill equation, the quality of the fit becomes poorer, but it still has a reasonable correspondence with the central part of the concentration range. We then see that the actual function of hemoglobin is best defined by a simulation that provides for separate, sequential binding of oxygen. But, the difference in the separate binding constants is not too large, and a very good approximation is obtained with the two state MWC model.

Monod, Wyman, and Changeux used the very good fit of their model to hemoglobin as support for the correctness of the concerted model for cooperativity.[23] We now see that this interpretation is, in fact, not completely correct. Given the dynamic nature of proteins, only rarely may four subunits in one tetramer ever behave in a concerted fashion. In the sense that it is the correct description for the binding behavior of hemoglobin, the MWC concerted model does not fit the oxygen binding data as well as the sequential model. But, it is worth noting that the MWC model still provides a quite satisfactory fit to the data. In the sense that many kinetic interpretations are made with the goal of some applied benefit, as in drug development, a simpler model for such binding experiments, as provided by the MWC equation, may still prove to be satisfactory.

6.3.1 Oxygen Binding Without Cooperativity

The numerous and well defined results showing positive cooperativity may lead to the conclusion that oxygen binding and conformational change (as shown in Fig. 6.2) are always linked. However, it is possible to demonstrate the binding of oxygen under conditions where external constraints prevent the tetramer from changing its conformation. Adult hemoglobin in either the deoxy- or oxy-hemoglobin conformation was enclosed within a silica gel, and oxygen binding was then measured.[185] When such preparations were made in aerated gels to stabilize the R conformation, binding was non-cooperative, but occurred with a high affinity. If the gels were made under anaerobic conditions to stabilize the T conformation, oxygen binding again occurred with no cooperativity, but with a very poor affinity. And, as noted in Chap. 6.2.2, when oxygen binding was measured with hemoglobin crystals, thereby constraining the conformational movement that occurs in solution, no cooperativity was evident.

Results such as these help to emphasize the importance of flexibility in proteins for dynamic activity. Oxygen binding was possible since the heme was available. Cooperativity did not occur, since movement for the rotation of one dimer relative to its associated dimer was constrained by the unique features of the hemoglobin's environment.

6.4 Hemoglobin is Also an Enzyme!

For much of the twentieth century hemoglobin was treated as an honorary enzyme. It was then known to have only the transport function, described above. However, since it was so well characterized, and therefore so useful in defining allosteric models which could also be applied to enzymes, it was almost uniformly included in descriptions of allosteric enzymes in most biochemistry textbooks.

More recent studies have now found results that show two different types of enzymatic activity by hemoglobin, in converting nitric oxide (NO) to nitrate, and also to an S-nitrosyl compound (S-NO).[165] NO can interact with the ferrous iron of the heme, or it can be oxidized by the oxygen at an occupied heme to form nitrate. The NO bound in the first step can then be transferred to cysteine-93 of one beta subunit, to form an S-nitroso-hemoglobin. This bound S-NO is then later released, to act as an effector on blood vessels. These three possible outcomes may be described schematically:

1. Oxidation $\quad Fe^{2+} - O - O + N = O \rightarrow [Fe^{3+} - OONO^-] \rightarrow Fe^{3+} + NO_3^-$

2. Addition $\quad Fe^{2+} + N = O \rightarrow NO - Fe^{2+}$

3. S-nitrosylation $\quad N = O + Cys\text{-}93 \rightarrow S\text{-}NO$

Reaction 1 occurs at hemes occupied by oxygen, while reaction 2 only occurs at empty hemes. Reaction 3 occurs only on cysteine-93, at the surface of beta subunits.[170] Although hemoglobins, even in the veins, are normally $\geq 70\%$ saturated with oxygen, reaction 2 (with unbound hemes) occurs with more hemoglobin molecules than reaction 1. This enzymatic conversion of NO to nitrate is a metabolic step in the catabolism of NO, but it appears likely that hemoglobin is not the only enzymatic path for this reaction. This is also the reaction that produces met-hemoglobin, in which the ferrous iron is oxidized to Fe^{3+}.

Reaction 3, S-nitrosylation, is important for normal blood flow, since the nitrosyl adduct forms more effectively with the R conformation (oxy-hemoglobin), in the lungs, and binds less well to the T conformation (deoxy-hemoglobin), and is being released as oxygen is unloaded in the microvasculature of tissues. This permits hemoglobin to control blood flow by the release of NO.

Blood flow is controlled by small sphincter muscles in the arteries that are activated by cGMP (made in response to appropriate physiological signals). The cGMP can be degraded to GMP by a guanylate cyclase, which must itself be activated by NO. The NO released by circulating erythrocytes is absorbed by the blood vessel epithelial cells, and binds to a guanyl cyclase receptor, leading to dilation of blood vessels and increased blood flow. Since the retention or release of S-NO by hemoglobin is influenced by whether it is in the deoxy state, then this mechanism provides a means for increasing blood flow according to the need (anoxic state) of the tissues. Thus, hemoglobin has two separate enzymatic reactions with NO that are not only relevant but perhaps necessary for the optimum delivery of oxygen to respiring tissues.

6.4.1 Summary of Hemoglobin Functions

A simple diagram for the passage of hemoglobin through the body's circulation is shown in Fig. 6.11. In the lungs the four subunits become fully oxygenated, and hemoglobin is in the R conformation. On reaching the capillaries, one or two O_2 may be released depending on the oxygen need of those tissues, which is expressed as the local pO_2. Conditions specific to the tissues may influence the unloading of oxygen. Carbon dioxide produced by energy metabolism is released by the tissues, and binds to the hemoglobin molecule. The binding of carbon dioxide near the N terminus of hemoglobin stabilizes the T conformation, which favors unloading of oxygen. Since carbon dioxide also produces local acidity, these protons will bind to the hemoglobin and help to change the conformation (the Bohr effect).

Hemoglobin returning in the veins is then partially deoxygenated, but also carries CO_2. Some of this CO_2 dissociates into bicarbonate plus protons. The remaining CO_2 is released in the lungs, to be exhaled. Proton pumps in the kidney continuously remove

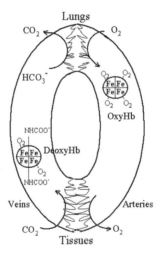

Fig. 6.11. The functions of hemoglobin as it circulates in the body. Oxygen is bound in the lungs and released to the tissues. In the tissues carbon dioxide is bound at the amino termini of the hemoglobin subunits and released to the plasma for buffering (HCO_3^-) or exhaled as a gas

protons, thereby increasing bicarbonate to concentrations normally above 25 mM. This provides mammals with an excellent buffering system, since acidosis is the most common acid–base disturbance. With the release of CO_2 in the lungs, the T conformation is not as stable, and loading of oxygen converts hemoglobin again to the R conformation, resulting in hemoglobin that is again fully loaded with oxygen.

One should appreciate the remarkable evolution of this protein that can accomplish multiple functions. It carries needed oxygen to the tissues. It removes the oxidative waste. It carries and enzymatically activates NO to serve as a signal molecule for control of the rate of blood flow. It releases NO when the pO_2 is low, thereby promoting better blood flow which increases the delivery of oxygen.

7

GLYCOGEN PHOSPHORYLASE

Summary

Glycogen, a polymer of glucose, is the major storage form for carbohydrates, and serves as a quick energy supply when it is converted to glucose-1-P, thereby stimulating glycolysis and oxidative phosphorylation. Glycogen phosphorylase is a central enzyme in controlling use of the stored glucose molecules. The process is regulated by hormones, via covalent modification of the enzyme by phosphorylation, as well as by normal intracellular effectors that define the energy state of the cell. The homodimeric enzyme has very large subunits, with three separate regulatory sites, plus the site for phosphorylation.

7.1 Overview of Glycogen Phosphorylase

After hemoglobin, in the historical development of enzymology, glycogen phosphorylase played an early and important role, as one of the first enzymes known to be allosterically regulated. As with hemoglobin, this was largely due to the ease with which it could be obtained. It occurs at high concentrations in liver, and accounts for about 6% of the soluble protein in muscle.[186] This made it easy to isolate and purify this enzyme from animal tissues, and permitted active research on the complex regulation of the enzyme.

Glycogen is used as an energy reserve. In the brain, it protects against hypoglycemia, since this organ functions constantly, while dietary intake of carbohydrates is intermittent. For muscles glycogen provides a more immediate, and a more abundant source of glucose-phosphate than does blood sugar. It thereby provides the precursors for ATP production that enable more rapid and more extensive muscle action for limited periods of required activity. It is therefore essential that cleavage of such stored glycogen molecules not occur steadily, since this would deplete the glycogen, or at best lead to a futile cycle as glycogen consumption balanced the synthesis of new glycogen.

The enzyme has therefore evolved to be sensitive to a number of different effectors that increase or decrease enzyme activity, appropriate to the metabolic state that leads to the effectors becoming more abundant. Glycogen phosphorylase was the first enzyme demonstrated to change activity after being phosphorylated.[187] This led to the current terminology for this enzyme: phosphorylase a is the active form, and phosphorylase b is the native enzyme, that is in the T conformation. Covalent modification of phosphorylase b, by the regulatory enzyme phosphorylase kinase (see Fig. 4.2) leads to the modified phosphorylase a, which is largely in the R conformation. In addition, as will be described below, each form of the enzyme is sensitive to a subset of regulatory effectors.

Interconversion between the two forms of glycogen phosphorylase is sometimes accompanied by a change in the oligomeric state. Phosphorylase b is more stable as a tetramer, and phosphorylase a as a dimer. Depending on the pH and temperature, phosphorylase a may also be tetrameric, but it has been concluded that the dissociation and association properties do not contribute significantly to the change in enzyme activity.[188]

Although it was observed fairly early that this enzyme has the cofactor pyridoxal phosphate (PLP), there has been extensive debate as to whether PLP participates in the chemistry of the reaction. The possibility that PLP served only a structural role was proposed with the discovery that PLP can be reduced with borohydride, with no change in enzyme activity.[186] Complete removal of PLP is accompanied by loss of activity, and also by the enzyme being unable to form a dimer, and remaining monomeric.[189] This suggested that in the evolution of this large multidomain enzyme, the PLP binding portion served largely a structural role.

However, there is good evidence that PLP is necessary for catalysis. The PLP binding site is close enough to the catalytic site for involvement of this cofactor to be possible. The phosphate of PLP has the appropriate pK of 6.1 that is consistent with the pH optimum for the reaction.[190] Early studies on borohydride reduction[188] appeared to eliminate a function for the cofactor, but a more systematic study has shown that only modification at carbon 4 of the pyridine ring is harmless. This is the site where borohydride reacts. Modification at the other positions of the ring leads to an inactive enzyme.[190]

7.1.1 Assay of Glycogen Phosphorylase

In vivo, the reaction catalyzed by the enzyme, termed glycogenolyis (7.1), is important since it produces glucose-1-P (Glc-1-P) from stored glycogen. This product is directly isomerized to glucose-6-P which enters the glycolytic pathway. Although plasma glucose, after entering a cell, must first be phosphorylated by use of ATP, the individual glucose residues on a glycogen chain are cleaved by phosphorolysis using inorganic phosphate, and produce the needed glucose-P without consuming ATP.

$$(\alpha\text{-}1,4\text{-glucoside})_n + P_i \rightleftarrows (\alpha\text{-}1,4\text{-glucoside})_{n-1} + \alpha\text{-D-glucose-1-P} \qquad (7.1)$$

At neutral pH the equilibrium constant for the reaction is 0.28, but under cellular conditions the reaction proceeds towards the production of glucose-1-P.[187] This is due to

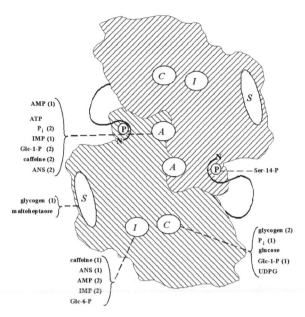

Fig. 7.1. Glycogen phosphorylase. A scheme for the dimer, showing the major binding sites: A (activator), C (catalytic), I (inhibitor), and S (storage). A number after a ligand's name designates the primary or secondary site for that ligand, based on the affinities (Table 7.1). The N-terminal peptide is normally disordered, but upon phosphorylation (at Ser14) folds onto the adjacent subunit. ANS, 8-anilino-1-napthalenesulfonate; UDPG, uridine-diphosphoglucose

the fact that cellular concentrations of P_i (8–10 mM) [31] are far greater than for glucose-1-P (about 170 μM). The reaction as written (7.1) is difficult to measure, since one product, $(\alpha\text{-}1,4\text{-glucoside})_{n-1}$, is indistinguishable from the substrate $(\alpha\text{-}1,4\text{-glucoside})_n$. The second product, glucose-1-P, is also not very easy to isolate from the other components in the assay mixture. Most assays of this enzyme therefore measured the reverse reaction, since it is very easy to quantitate the formation of inorganic phosphate by a very specific colorimetric assay.

7.2 Regulation of Glycogen Phosphorylase

Glycogen phosphorylase has very large subunits, with an M_r of 97,444 Da. This makes it truly large compared to most enzymes (Fig. 1.5), and this extra size provides extra binding sites. The various ligand binding sites of glycogen phosphorylase are diagrammed in Fig. 7.1. The enzyme is normally a dimer, and the A sites, for binding the activator AMP, are at the dimer interface. The catalytic site, C, is in the center of the subunit, and an inhibitory site, I, is 10 Å away. A storage site, S, permits the enzyme to attach to a large glycogen complex. This feature has an evident benefit. Due to its enormous size glycogen is not very soluble. It is therefore possible for many enzyme molecules to attach to a single glycogen aggregate, and this helps minimize the time for

Table 7.1. Effects of ligand binding at different sites on glycogen phosphorylase[a]

Ligand	R conformation			T conformation		
	Catalytic Site	Activator Site	Storage Site	Catalytic Site	Activator Site	Inhibitor Site
Catalytic site						
Glycogen	↑22 mM		↑ 0.5 mM			
P$_i$	↑				↓	
Glc-1-P	↓ 4.7 mM				↓	
Glc-6-P	↓20 µM					
Glucose				↓ 0.8 mM		↓ 1.7 mM
UDPG	↓1.4 mM					
Activator site						
AMP		↑2 µM			↑ 25 µM	
ATP					↓	
IMP		↑50 µM				↓ 4 mM
ANS		↑140 µM				↓ 3 mM
Inhibitor site						
Caffeine						↓ 80 µM
Purine						↓

[a]Compounds are listed under their primary binding site, though many bind at additional sites. Arrows indicate the effect of binding on enzyme activity, for the physiological reaction. However, Glc-1-P is actually the substrate for the in vitro assay. Some K_d values are shown, from Fletterick and Madsen,[188] Madsen et al.[192] Cori and Cori,[193] and Sprang et al.[194]

the encounter between enzyme and substrate. It has been observed that for the reverse reaction, binding at the storage site is necessary for optimum activity,[193] and malto-heptaose is frequently used to serve this function during in vitro assays. In addition to the four binding sites for ligands, there is a fifth site at which the enzyme can be covalently phosphorylated. This occurs near the N terminus, at Serine-14.

In an in vitro assay, many of the normal ligands bind at two of the sites on the enzyme, often with opposing consequences. However the binding affinities are usually very different, so that a large range of ligand concentration is necessary to achieve binding at both sites. It is not as likely that in the cell the same range of concentrations for each ligand will occur. Under normal conditions in vivo, for most ligands the primary binding site is the only relevant site, while glycogen, however, must bind at both sites shown. To help distinguish differences in affinity, numbers after the name of the ligand define the primary and secondary binding sites. These are based on the known affinities (Table 7.1).

7.2.1 Two Different Activating Mechanisms for Two Different Physiological Needs

There are two separate mechanisms for activating the enzyme:

1. Depletion of ATP levels leads to increased AMP concentrations, which activate the enzyme and results in higher ATP synthesis.
2. The organism has a need for maximum skeletal muscle activity (fight or flight); this produces a hormonal signal that results in covalent phosphorylation of the enzyme, which produces the maximum enzyme activity, and results in higher ATP concentrations.

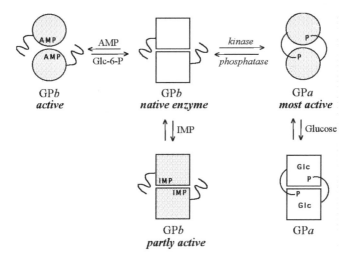

Fig. 7.2. Allosteric regulation of glycogen phosphorylase. *Circles* are *R* conformers, *boxes* are *T* conformers; shaded forms denote the more active species. Phosphorylase *b* is the native enzyme. It may be activated by binding AMP, or converted to phosphorylase *a*, by being itself phosphorylated. IMP, inosine-monophosphate

This represents a very clever division of the signaling for the two distinct regulatory strategies. If internal cellular ATP levels are less than optimum, the consumption of ATP has produced increasing concentrations of AMP. A binding site for AMP that produces a conformational change with greater activity is a logical property for this enzyme (Fig. 7.1), since by activating glycogenolysis, it favors restoring the ATP levels in that cell. Since AMP is consumed in making ATP, AMP concentrations will decline, as ATP concentrations are increased. The enzyme activity will naturally respond by returning to a lower level of activity.

If the whole organism has a sudden need for accelerated muscle function, then a hormonal signal will reach all skeletal muscle cells. Because this is a more urgent situation, the hormonal signal results in an intracellular phosphorylation of glycogen phosphorylase. When covalently modified, the enzyme is for a limited time maintained in this optimum conformation, leading to the maximum production of Glc-1-P and ATP. The time frame for this altered state is normally from many minutes up to about an hour, as the protein phosphatase steadily returns enzymes to their native state (Fig. 4.2). At this time glycogen usage would again be minimal, and the organism would return to using circulating blood glucose. Figure 7.2 summarizes the conformational changes of the enzyme, with the related change in affinity.

7.2.2 Ligands Binding at Two Different Sites on One Subunit

Early kinetic studies with the enzyme showed that phosphorylase *b* was in the *T* conformation, as defined by cooperative binding of the substrate Glc-1-P. This is evident in the lowest curve of Fig. 7.3A, where the activator AMP had been added at 52 μM, so that the A site is not saturated, and the enzyme population would then be a mixture of

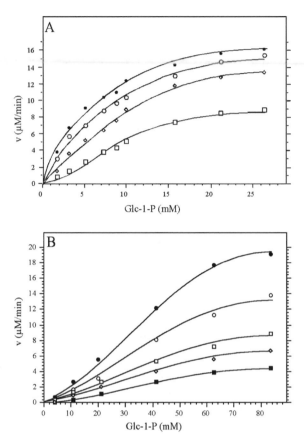

Fig. 7.3. Activity of glycogen phosphorylase. (**A**) in the presence of AMP at concentrations of (*open square*) 52 μM, (open *diamond*) 104 μM, (*open circle*) 208 μM, (*filled circle*) 2.08 mM; (**B**) in the presence of IMP at concentrations of (*filled square*) 8.3 μM, (*closed diamond*) 33.3 μM, (*open square*) 83 μM, (*open circle*) 166 μM, and (*filled circle*) 416 μM. Note the change in the substrate concentration range. Data from Black and Wang[195]

T and *R* conformations. As the concentration of AMP was increased to 104 μM, there remains only a suggestion of cooperativity, and the curves become hyperbolic, indicative of constant affinity, which would result as the enzyme population is completely converted to the *R* conformation.*

---

*
  The concentrations stated for Fig. 7.3A are inconsistent with affinity values for AMP found by later researchers (Table 7.1), since the enzyme should have been saturated at the lowest AMP concentration used. Perhaps the AMP used was not pure, or had not been properly quantitated. These are the most detailed kinetic studies for the effect of AMP and IMP.

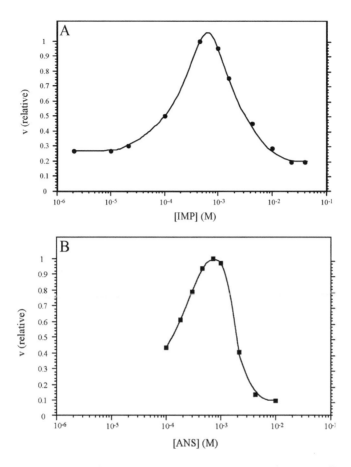

Fig. 7.4. Affinity measurements for IMP (inosine-5′-monophosphate) and ANS (8-anilino-1-napthalene-sulfonate), as determined by the change in the enzyme's activity. Data from Madsen et al.[192]

By comparison, the other activator, IMP (produced by deamination of AMP), produces an activation effect, but is not able to convert the enzyme population completely to the *R* conformation. This is evident by the curves remaining sigmoidal, even at high concentrations of IMP. Crystal structures obtained later with the enzyme binding IMP showed that IMP binds very effectively at the I site on the enzyme. If the binding constants for the two sites are not sufficiently different, then before the ligand can saturate the high activity binding site, it will begin to bind at the second site, and the kinetic activity profile will be a summation of activity due to binding at the separate sites.

This is made more explicit with examples of ligands binding at completely different sites on this enzyme (Fig. 7.4). Comparable to the more important activator AMP, IMP also binds at the A site, causing an increase in enzyme activity (Fig. 7.4A). The activity curve was measured over an IMP concentration range of over 4 logs. This is much more extensive than the concentration range used in most enzyme studies. We see an increase

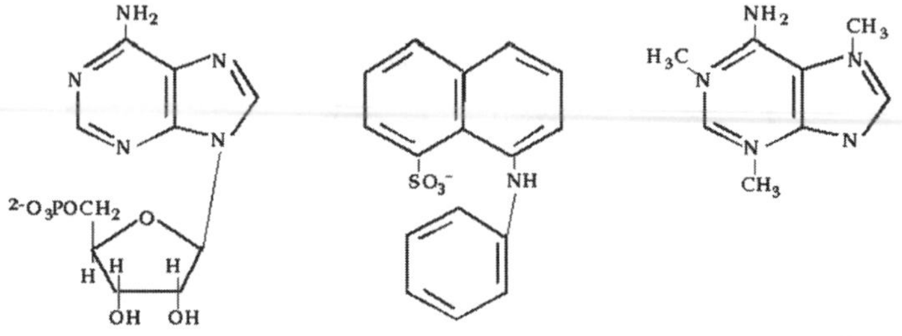

Fig. 7.5. Structures for AMP and purine analogs. Caffeine is a trimethyl purine, and mimics the purine base

in enzyme activity as IMP increases 100-fold, from 10^{-5}M to 10^{-3}M. This is similar to a normal 2-log response. After reaching this peak in activity, it is followed by a decrease in activity as IMP is increased an additional 20-fold. This pattern reflects tight binding at the *A* site (K_d = 50 μM), with activation, followed by weaker binding at the I site (K_d = 4 mM), with inhibition. Although binding at the I site is weaker, when the enzyme becomes saturated with IMP it adopts a new conformation that becomes the dominant form. The structural linkage between the A site and the I site permits two alternate conformations, and only one site at a time may be occupied on one enzyme subunit. In a mixture of phosphorylase molecules, some will be in the *T* conformation, and then will be stabilized by ligands binding at the I site. Other enzyme molecules will be in the *R* conformation, and then will be stabilized by ligands binding at the A site. The activity curve observed always represents the sum of such a mixture of conformations.

Under physiological conditions, IMP is very unlikely to ever exceed a concentration of 1 mM.[31] Then, within a cell, IMP acts only as an activator, although weaker than AMP, because its affinity at the A site (K_A = 50 μM) is much poorer than AMP's (K_A = 2 μM) as shown in Table 7.1. We see a similar response pattern by the enzyme in the presence of varying concentrations of ANS (8-anilino-1-napthalenesulfonate). The structures for AMP, caffeine, and ANS (Fig. 7.5) are of comparable size and stereochemistry, and we see that they may compete for similar binding sites. Caffeine is a trimethyl purine, and ANS has two comparable hydrophobic rings, and the attached sulfate has a charge and size similar to the phosphate on AMP and IMP. The binding curve for ANS is not as extensive (Fig. 7.4B), but it has the same pattern of first activating and then inhibiting the enzyme. Again, this is logically explained by ANS binding at two different sites on the enzyme. Results such as those shown (Fig. 7.4) have led to the designations for primary and secondary sites as noted in Fig. 7.1.

7.2.3 Activation by AMP and IMP

A simple overview of the various conformational states is shown in Fig. 7.2. Under normal conditions, the enzyme exists mostly as glycogen phosphorylase *b*. The enzyme in the *T* conformation has very little activity, due to poor affinity for the substrates. The major intracellular effector is AMP, since this molecule becomes more abundant as ATP is depleted by muscle action. Binding of AMP to phosphorylase *b* leads to a more active form of glycogen phosphorylase b, denoted by the conformational change. The A site at which AMP binds may also bind IMP, and both effectors increase the affinity for Glc-1-P, the substrate used in the normal assay. IMP has a poorer affinity than AMP at the A site. Although IMP also binds weakly at the I site, the concentration required means that this cannot be significant in vivo.

The A site is located at the interface of the two subunits in the dimer (Fig. 7.3), and the binding of AMP requires contact with the adjacent subunit,[196] facilitating the cooperative binding of AMP.

7.2.4 Activation by Other Metabolic Effectors

Two other molecules also activate the enzyme. These are the two physiological substrates, glycogen and inorganic phosphate. The activating effect of glycogen is caused by binding at the storage site, since glycogen, or a glycogen substitute such as maltoheptaose, is required for maximum enzyme activity. P_i binds at the catalytic site as a normal substrate, and also at the A site, where it fits the phosphate portion of the AMP binding site. At the A site P_i acts as an incomplete AMP. This interpretation is based on results with sulfate, since this compound is seen at the A site in crystal structures, and also produces a modest activating effect in kinetic studies.[197] P_i is not large enough to produce as strong a conformational change as is caused by AMP.

7.2.5 Activation by Phosphorylation

An early and important discovery was that the enzyme could be regulated by hormonal control,[193] and this is diagrammed in Fig. 4.2. An increase in circulating epinephrine increases hormone binding to a membrane receptor, which activates an intracellular adenyl cyclase, resulting in production of cAMP. This cyclic nucleotide is the effector that activates a protein kinase, and thereby initiates the cascade of kinases resulting in the phosphorylation of glycogen phosphorylase b. When this protein is itself phosphorylated, by phosphorylase kinase, it becomes the more active enzyme, glycogen phosphorylase a. This is the fully active R conformation.

As diagrammed in Figs. 7.1 and 7.2, the phosphorylation site is at Serine-14, near the N terminus. This segment of the protein is normally disordered, because it is not visible in crystal structures of phosphorylase b. Upon covalent phosphorylation, this peptide segment assumes an ordered shape, and interacts with residues in the adjacent subunit. The position of this phosphate is near the position of the AMP, when bound at the A site. Both groups appear to have a common set of conformational switches, leading to the same active conformation in the R state.[196] When glycogen phosphorylase b, with AMP bound, is phosphorylated by phosphorylase kinase, there is some additional increase in activity. This reflects the fact that covalent modification affects a greater number of the subunits than does AMP binding. Since binding is transient, it is difficult to fully saturate the A site by AMP.

While the two activating signals may appear to duplicate each other, they are the result of different stimuli. We also see that the hormonal signal, which represents a state of urgency, results in the maximum activity by glycogen phosphorylase. And, this activity will be maintained as long as the covalent phosphate remains attached. Unless the hormonal signal is itself maintained, the protein phosphatase will steadily dephosphorylate glycogen phosphorylase *a*, thereby converting it back to glycogen phosphorylase *b* (see Fig. 4.2). The hormonal signal therefore results in a somewhat stronger increase in activity, and this will not begin to decline as AMP is being consumed for the synthesis of ATP.

7.2.6 Summary of Effector Binding

The above discussion has focused on activator ligands binding at the A site, and inhibitor ligands binding at the I site. Careful inspection of Fig. 7.1 combined with the effects on activity noted in Table 7.1, show that a ligand may be inhibitory even though it binds at the A site. AMP is the normal obligatory activator. Compounds that sufficiently resemble AMP will bind at the A site and produce a comparable activation effect. However, compounds that contain only part of the features involved in binding by AMP, may then bind incorrectly at the A site. This will not produce activation, since that requires correct binding that can stabilize the *R* conformation. Compounds that bind incorrectly at the A site will then compete against AMP for binding, and thereby lead to reduced activation by AMP itself, which will be observed as inhibition. An example of such a compound is Glc-1-P. In contrast, the isomerized Glc-6-P appears not to bind significantly at the A site, and causes only inhibition at the catalytic site.

An unresolved feature of this complex enzyme is the normal physiological function of the I site. After 70 years no tight binding physiological compound has been identified. Several ligands that bind at the I site are shown in Table 7.1. Though caffeine has a good affinity, it is not a normal cellular compound. The other inhibitors listed have poor affinity, so that inhibition may be demonstrated in vitro, but the cellular concentration of these compounds would not permit them to be significant inhibitors under normal conditions.

7.3 Evolution of Glycogen Phosphorylases

Most phosphorylases from organisms that have been studied show at least one of the regulatory strategies described earlier (Table 7.2). Bacteria only have activation by AMP at the A site.[198] The phosphorylation site is a later addition with the appearance of eukaryotes. We see some evidence that in unicellular eukaryotes both strategies were not always maintained. Yeast show the ability to activate glycogen phosphorylase by phosphorylation, but appear to have lost the ability to be activated by AMP.[197] The plant phosphorylase, from potato, gives no evidence of either form of regulation.

The hormonal control described above for mammalian glycogen phosphorylase is a logical development for an active multicellular organism. Activation by AMP, at the A site, as found in bacteria, would continue to be beneficial for regulating the energy supply within a muscle cell. As AMP concentrations increased, due to the depletion of ATP, this would activate glycogenolysis and result in greater ATP production, thereby consuming the AMP, and returning the enzyme to the normal (less active) state. If the organism had a sudden need for dynamic activity (fight or flight situation), the crisis would cause release of epinephrine. Binding of this hormone to muscle cell receptors (Fig. 4.2) would activate the intracellular protein kinase cascade, and thereby produce phosphorylation of

Table 7.2. Regulatory properties of glycogen phosphorylase from different organisms[a]

Species	Activated by AMP	Activated by phosphorylation
E. coli	Partly; increase in V_{max}	No
Potato	No	No
Yeast	No	Yes
Rabbit muscle	Yes; up to 80% of V_{max}	Yes

[a]From Newgard et al.[199]

glycogen phosphorylase *b*. The resulting phosphorylase *a* has the maximum activity, and the covalent change will last from many minutes up to an hour, until the phosphorylation is undone by the action of the protein phosphatase. Thus the hormonal signal produces a more optimum response, and guarantees the maximum production of Glc-1-P, to maintain both oxidative phosphorylation as well as anaerobic glycolysis at peak levels.

Although most plants are multicellular, their glycogen phosphorylase is not regulated by phosphorylation. Presumably plants, which have no need for rapid action, have not inherited or evolved this form of control. Plants also store more carbohydrates as starch, which is cleaved by amylase. They therefore have much larger additional supplies of stored glucose that they may utilize. Possibly for this reason they have not maintained the strategy for activation of their glycogen phosphorylase by AMP.

7.3.1 Isozymes of Glycogen Phosphorylase

The liver stores glycogen as a glucose reservoir for the whole body, though this is largely destined for the central nervous system during times of hypoglycemia. Table 7.3 defines the functions for the different mammalian isozymes, and also lists the most important form of activation for each isozyme (e.g., phosphorylation in muscle, but AMP activation in brain). Liver metabolizes fatty acids routinely, and can obtain all its energy form the oxidation of fats. Therefore liver does not routinely need to activate the enzyme for intracellular energy needs, and therefore this isozyme has the weakest K_A for AMP (Table 7.4). But, liver needs to be responsive to hormones that signal the need for glucose in the blood, such as glucagon, a signal for hypoglycemia, and epinephrine, a signal for rapid muscle action. Whenever the muscle phosphorylase is activated by a hormonal signal, this will equally activate the liver enzyme. Therefore the liver isozyme is very sensitive to activation by phosphorylation alone. Since the hormonal signal functions to produce glucose for the blood, glucose itself is the most effective inhibitor.

Muscle also stores glycogen, but is not able to convert the Glc-1-P produced to free glucose, and uses the stored glycogen exclusively in the muscles cells for muscle action.

Table 7.3. Regulation of mammalian isozymes[a]

Type	Occurrence	Function	Regulatory controls
Muscle	Muscle	Generate glucose-P for extra muscle activity	+ : phosphorylation, AMP − : glucose
Liver	Liver, Peripheral tissues	Generate glucose to export for other tissues	+ : phosphorylation − : glucose
Brain	Fetal tissues; Adult brain	Generate glucose-P during anoxia or hypoglycemia	+ : AMP, phosphorylation − : glucose

[a]Human tissues normally have one isozyme in abundance, but also contain the other two[199]

Table 7.4. Kinetics of mammalian isozymes[a]

Isozyme	Basic activity (%)	K_A AMP (μM)	n_H
Muscle	0.3	60	1.9
Liver	40	75	1
Brain	9	9	1

[a]From Guénard et al.[200]

When muscle is resting, it can obtain adequate energy from the use of fatty acids, and glycogen consumption is reserved for active work. Under conditions of active work the muscle phosphorylase is then mainly activated by phosphorylation, in response to a hormone signal, and is also activated by AMP binding at the A site. Glucose itself remains the most effective inhibitor.

Brain cells function constantly. With a mass of about 1.4 kg, the brain normally represents 2% of an adult body weight of 70 kg. Yet the brain is so active that it consumes 20% of the oxygen used under resting conditions. This energy demand cannot be satisfied by the use of fatty acids, and explains the need for glucose as the chief energy source. Brain has only a modest amount of glycogen, and this is a temporary reserve for very brief periods of anoxia, or periods of hypoglycemia. It is therefore logical that the brain isozyme is mainly activated by AMP, in response to the intracellular energy state.

These important differences in tissue function require differences in the control of the enzyme. And the most important kinetic values are summarized in Table 7.4. Note that although AMP is the activator for all three isozymes, a benefit of having isozymes is their individual change in the basal activity, and sensitivity to AMP. The muscle enzyme is to a greater extent in the T conformation, under resting conditions, and has the lowest normal activity. Therefore it also shows cooperativity for the substrate. Because activation by AMP is not as important, it has a poorer affinity for this nucleotide.

The liver enzyme has the highest average basal activity, since hypoglycemia begins within 1–2 h after a meal, and the liver must then replenish blood glucose. Because the liver itself may obtain adequate energy from fatty acids, the liver has limited needs for using carbohydrates to produce its energy. Therefore this isozyme no longer shows significant dependence on AMP. This is consistent with the liver isozyme also having the poorest affinity for AMP. In contrast, the brain enzyme has the highest affinity for AMP, so that if brain ATP concentrations decline even modestly, the brain isozyme is pro-portionately more activated.

8

PHOSPHOFRUCTOKINASE

Summary

Phosphofructokinase is the principal regulatory enzyme for glycolysis. Gene duplication plus fusion have doubled the size of the eukaryotic enzyme. The smaller enzyme in microbes has versions that may use ATP, ADP, or pyrophosphate as the phosphate donor. These enzymes are normally inhibited by phosphoenolpyruvate and activated by AMP. The larger enzyme uses ATP, and has additional regulatory sites. A special site for fructose-2,6-bisphosphate has resulted in this activator being obligatory for many versions of the enzyme. The larger enzyme is also activated by ADP/AMP, and inhibited by ATP, and citrate.

8.1 The Diversity of Phosphofructokinases

The common feature of phosphofructokinases (PFKs) is their physiological role in synthesizing fructose-1,6-P_2 from fructose-6-P. The great majority of all kinases use ATP as the donor for the high energy phosphate to be transferred to the acceptor. Naturally, there is then a family of PFKs that uses ATP. However, there also exists a family that uses ADP, and a third group that functions with pyrophosphate (PP_i) (8.1). For simplicity, these will be referred to as ATP-PFKs, ADP-PFKs, and PP_i-PFKs. Although enzymes of each group share more features with other members of that group, the many common features of all these enzymes support the model of a single common ancestor.[201, 202] Features that are shared by all three groups, as well as distinguishing features are shown in Table 8.1.

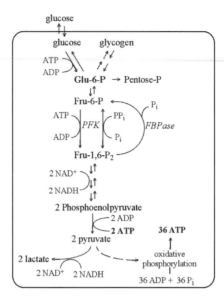

Fig. 8.1. Glucose metabolism. A simplified scheme to emphasize phosphorylation steps and redox steps. PFK, phosphofructokinae; FBPase, fructose bisphosphatase

$$ATP + Fru\text{-}6\text{-}P \rightarrow Fru\text{-}1,6\text{-}P_2 + ADP$$
$$ADP + Fru\text{-}6\text{-}P \rightarrow Fru\text{-}1,6\text{-}P_2 + AMP \qquad (8.1)$$
$$PP_i + Fru\text{-}6\text{-}P \rightarrow Fru\text{-}1,6\text{-}P_2 + P_i$$

The early position of phosphofructokinase in the glycolytic pathway (Fig. 8.1) makes it an obvious candidate for significant regulatory control in those organisms or tissues where the energy need may also be supplied by oxidation of fatty acids under normal conditions. For complex organisms, such as mammals, this permits carbohydrates to be stored, and then used preferentially in those tissues with a greater glucose requirement. In this category are erythrocytes and the brain. Erythrocytes are unusual cells, in that the mature cells no longer possess mitochondria or nuclei. This structural simplification permits them to contain greater amounts of hemoglobin within each cell. It also means that oxidative phosphorylation is not possible, so that anaerobic glycolysis is the only mechanism for generating ATP. Brain cells do have mitochondria. But, the brain functions at a constant pace day and night; and its energy demands are better maintained by a larger dependence on glucose metabolism. Therefore, some of the regulatory constraints described for glycogen phosphorylase (Chap. 7) apply also to phosphofructo-kinase in complex organisms.

For microbes, the need to regulate phosphofructokinase may simply be a cycling between the available fuel sources. When fats or amino acids are more abundant, any stored carbohydrates can be spared. As seen in Fig. 8.1, some mammalian tissues are gluconeogenic, and can convert 3-carbon carbohydrates to 6-carbon sugars. While many glycolytic reactions are easily reversible, at the step mediated by phosphofructokinase in

Table 8.1. The different types of phosphofructokinases

Type (species)	Subunit size (kDa)	Oligomer size	$K_{0.5}$ (μM)[a] F-6-P	K_m (μM) P donor	k_{cat} (s^{-1})	pH optimum	Refs.
ATP-PFKs							
E. coli	37	α_4	25,000; 12.5	60	0.12		203
Yeast	108; 104.6	$\alpha_4\beta_4$	900	48	0.03		204
D. discoidium	92.4	α_4	22	16	0.9	7.6	205
Rat	82	α_4	9	42	180	8	206
ADP-PFKs							
P. furiosus	52	α_4	2,300	110	1.7	6.5	207
PP$_i$-PFKs							
S. tuberosum	67; 57	$\alpha_2\beta_2$	320	70	0.1	7.3	202
E. gracilis	110	α	300	100	0.2	6.6	208, 209

[a]Where two values are shown, the first is for the *T* state (absence of effectors), the second for the *R* state

the glycolytic path, an opposing enzyme, fructose bisphosphatase (FBPase), is required for the gluconeogenic reaction. And, having separate enzymes for these two opposing reactions permits each enzyme to be independently expressed at differing rates, and also independently regulated. But, microbes do not appear to need such sophisticated controls, and we will see that the enzyme PP$_i$-PFK is easily reversible, and combines the reactions mediated by the normal ATP-PFK and FBPase in other organisms.

8.1.1 Summary of Phosphofructokinases

Table 8.1 lists a few examples for the three major types of phosphofructokinases. Inspection of this table shows a considerable range for k_{cat}. In Chap. 3, I discussed the experimental support that a k_{cat} of 1 s^{-1} is a lower limit for metabolic enzymes. Since several enzymes in Table 8.1 have values lower than this, it is likely that they were not completely pure, or that they were assayed under less than optimal conditions.* Note that the rat enzyme has a very normal activity for a kinase (see Fig. 2.4).

The extended range for the size of subunits for this enzyme results from an expansion in the ancestral enzyme by gene duplication, and will be described below. The ancestral enzyme had a subunit size below 40 kDa. Consistent with the discussion in Chap. 1, larger size commonly correlates with extra binding sites on an enzyme subunit. We also see examples for gene duplication in several species, since the functioning oligomer contains two types of isozymes as subunits. The well studied enzymes from *E. coli* and rat are both homotetramers and allosteric. Therefore the $\alpha_4\beta_4$ heterooligomers (Table 8.1) are an example of an alternate strategy to arrive at an optimum allosteric system, comparable to the mammalian hemoglobin $\alpha_2\beta_2$ tetramer.

Interesting data ranges in the table are for the values of the affinities for the two substrates. The K_m values for the phosphate donors are in the range at $\leq$100 μM,

---

*The value for k_{cat} is sensitive to the enzyme's purity, since it represents the observed activity per unit of protein. Affinity values are more likely to be the same for pure and partially pure enzyme samples.

suggesting moderate binding. This value may reflect the need for the enzyme to be at least half active, in response to available ATP, when that concentration of ATP has become limiting for general metabolism. A possible outlier is the enzyme from *D. discoidium*, but no explanation for the higher affinity has been found.

By comparison, we see a thousandfold range in the affinities for fructose-6-P. This may simply represent the difference in the affinity of the *T* state (very poor affinity), and the *R* state (very good affinity). This difference was specifically measured by Blangy et al. for the *E. coli* enzyme,[203] and we see that the change in the enzyme's conformation leads to a 2,000-fold greater affinity for fructose-6-P. To have the enzyme predominantly in the *R* conformation, it is necessary to include one of the activators for this enzyme in the activity assay. While this is frequently done, authors have not been consistent in describing the conditions under which the K_m values were determined.

Support for the above distinction of high affinity values being for the *R* state comes from data for the enzyme from the slime mold, *D. discoidium*. This enzyme is unusual for phosphofructokinases; it has not demonstrated any significant allosteric features.[205] This means that this enzyme is completely, or mostly, in the *R* conformation,[†] and the affinity for fructose-6-P found for this enzyme is very high, consistent with the enzyme being in the active *R* conformation.

8.1.2 Properties and Emergence of Phosphofructokinases

All three types of phosphofructokinase are found in bacteria. The emergence of the ADP-PFKs and the PP_i-PFKs may represent opportunistic adaptations for microbes in special environmental niches. For most organisms, ADP is the product that is obtained when ATP is used in some reaction. The gamma phosphate bond of ATP has the highest energy, so that reactions using this group in phosphotransfer are generally not reversible under cellular conditions. However, the beta phosphate bond is sufficiently energetic that, in the thermodynamic sense, it should still be a good reagent in many phosphotransfer reactions. Since the reverse reaction, where ADP had been the substrate, is energetically less difficult, phosphotransfer reactions from ADP as the initial phosphate donor are more reversible (8.2).

$$\begin{array}{lll} ATP + X \rightleftarrows ADP + \text{X-P} & \text{Irreversible} & \\ ADP + X \rightleftarrows AMP + \text{X-P} & \text{Modestly reversible} & (8.2) \\ PP_i + X \rightleftarrows P_i + \text{X-P} & \text{Reversible} & \end{array}$$

Evolution thus presented organisms an interesting choice: the organism may obtain an additional chemical reaction by using ADP in place of ATP, but it does this by losing some

---

[†] Allosteric features, as manifested by positive cooperativity, become evident only when there is enough of a change in L, the ratio of $[E_T]/[E_R]$.

of the specificity and control for that particular reaction step, since the reaction is now modestly reversible. This logic then also applies to the use of pyrophosphate as the phosphate donor, since transfer reactions involving pyrophosphate are nearly isoenergetic, and quite easily reversible. If, for any microbe, the most stringent need is survival with the minimal available energy, then using an ADP-PFK or a PP_i-PFK would be beneficial. If the ability to control flux down the glycolytic pathway is more important, due to the availability of alternate energy sources, and the benefit of being able to choose between/among them, then an ATP-PFK is the more advantageous.

Additional constraints may also influence the choice of the type of phospho-fructokinase. The extreme thermophile *P. furiosus* has an ADP-PFK. It is possible that at temperatures near 90°C, ATP is less stable and becomes hydrolyzed to ADP at some constant level. This would make ADP concentrations more stable, and potentially a better source for phosphotransfer reactions.

The benefit of metabolic control comes at an energy cost. Higher, more complex organisms, with steady energy from photosynthesis (plants) or available foods (vertebrates), have benefited from a larger energy budget, and give evidence of this in having greater control of their metabolism.[210] ATP-PFK is found in most of these organisms, with one exception listed in the table for *S. tuberosum* using the PP_i-PFK. However, since the bulk of this plant exists as an underground tuber, this may be a logical adaptation.

8.1.3 PP_i-PFKs

The importance of gluconeogenesis was described in Chap. 7. The existence of the fairly irreversible ATP-PFK in most organisms makes a second enzyme, fructose-1,6-bisphosphatase (FBPase) necessary to convert gluconeogenically produced fructose-1,6-P_2 to fructose-6-P, so that this sequence may continue toward the formation of glucose from appropriate precursors, such as amino acids. If, for some organism detailed control is not as necessary, then the two opposing enzymes, PFK and FBPase, may be replaced by the single enzyme PP_i-PFK (Fig. 8.1). This would provide some economy, since only a single enzyme must be synthesized, and also since the lower energy in the pyrophosphate bond still may produce a metabolite (fructose-1,6-P_2) that will, a few steps later, lead to ATP synthesis (Fig. 8.1).

Some microbes express both an ATP-PFK and a PP_i-PFK, but under different nutritional conditions. As an example, the actinomycete *Amycolatopsis methanolica* can survive on 1-C compounds, such as methanol and formaldehyde. When grown on glucose the fungus expresses only the PP_i-PFK.[211] But when grown on methanol or formaldehyde as the carbon source, they have only very low levels of PP_i-PFK, and now express very high levels of ATP-PFK. These organisms have the ability to use 1-carbon compounds, by incorporating them into ribulose-5-P, leading to the formation of fructose-6-P. It is not yet clear why the pathway of fructose-6-P formation should correspond to a different type of phosphofructokinase.

8.2 Regulation of Phosphofructokinase

8.2.1 Regulation of the Phosphofructokinase in *E. coli*

As one of the authors of the MWC model, Jaques Monod chose the phosphofructokinase from *E. coli* for initial kinetic studies to evaluate his model with this allosteric enzyme. In their studies, the enzyme was not completely pure, but their enzyme preparation proved to be quite stable, and produced very reproducible kinetic values.[203] One set of their studies illustrated the importance of the activator, ADP (Fig. 8.2). Without this activator, the enzyme is in the *T* conformation, and therefore has a very high K_m for fructose-6-P, at 25 mM (Table 8.1). Although the experimental concentrations of fructose-6-P originally

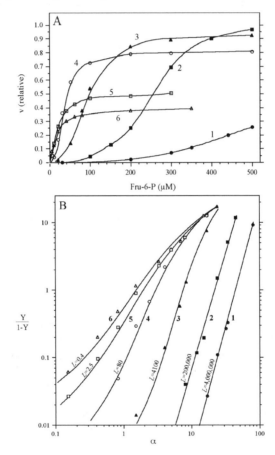

Fig. 8.2. The activity of phosphofructokinase from *E. coli*. For the six curves, ADP concentrations were: #1, 0; #2, 20 μM; #3, 70 μM; #4, 220 μM; #5, 520 μM; #6, 820 μM. (**A**) direct kinetic plot; (**B**) Hill plot for the same data, from Blangy et al.[203] $\alpha = [\text{Fru-6-P}]/(1.25 \times 10^{-5} \text{ M})$

went up to 500 µM, this results in very low activity (curve #1), showing that this substrate is not very effective in promoting the R conformation. Addition of ADP, first at 20 µM (curve #2), then at 70 µM (curve #3), clearly produces much better activity. The K_A for ADP is 20 µM, and at this concentration only half the enzyme molecules should be in the more active R conformation.

We naturally expect that further increasing the concentration of the activator should produce much greater activity. When ADP was increased to 70 µM, the maximum activity did not change. But, since curve #3 is shifted further to the left, we see evidence of a greater shift to the R conformation, which has the lower K_m of 12.5 µM. One interpretation for the mixture of results in curve #3 (better affinity, but no increase in maximum activity) is that ADP is now binding at a regulatory site to stabilize the R conformation, and also binding at the catalytic site as an inhibitor against ATP. This was supported by studies with GTP and GDP. GTP is less effective as a phosphate donor compared to ATP. The K_m for GTP is 1.2 mM, for ATP it is 60 µM. Monod et al. observed that GDP is an equally effective activator. Since it is not an effective inhibitor, only the increase in activity was observed.

Because the kinetic curves (Fig. 8.2A) show positive cooperativity for fructose-6-P, this substrate itself can stabilize the R conformation. However, the affinity, even for the R conformation is not strong enough. Although no value is available for $E.$ $coli,$ we know that the average concentration for this substrate in rat heart is only about 4 µM (Table 1.4). If we assume that bacterial concentrations for this metabolite are comparable to those in rat, then less than 30% of the enzyme can ever be stabilized in the R conformation due to this substrate alone. This is consistent with the values for curve #1 (Fig. 8.2A). We then understand that while the enzyme may be regulated by its substrate, fructose-6-P, in vitro, this allosteric response is not sufficient. Within cells some additional activator is necessary, and this role is performed by ADP in most microbes, and by AMP in most eukaryotes.

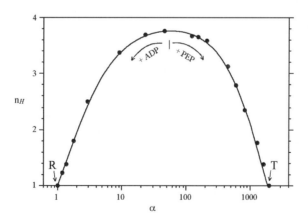

Fig. 8.3. Hill plot for binding of fructose-6-P by $E.$ $coli$ phosphofructokinase. Data from Blangy et al.[203]

The maximum cooperativity for fructose-6-P is for the T conformation, in the absence of effectors. This represents the change in L, as the R conformation becomes more abundant, while it is stabilized by increasing concentrations of fructose-6-P (Fig. 8.2b). This may also be diagrammed as a plot of the Hill coefficient vs. α, the normalized substrate concentration (Fig. 8.3). The enzyme is a tetramer, and n_H has a maximum value of 3.8. This demonstrates how remarkably cooperative this example of phosphofructokinase has become. With the addition of ADP, the enzyme becomes increasingly stabilized in the R conformation with a lower value for K_m (12.5 μM). In the presence of the inhibitor PEP, the enzyme becomes increasingly stabilized in the T conformation, and K_m becomes much larger (25 mM).

It should be noted that the end points of the curve in Fig. 8.3 define the complete R conformation (due to high concentrations of ADP) and the complete T conformation (due to high concentrations of PEP). Although the free enzyme, which is greater than 99% in the T conformation, has a very high n_H, the presence of ligands to stabilize either conformation has the effect of freezing the enzyme in that conformation, as shown by the Hill coefficient remaining at 1.0. This is comparable to the studies with hemoglobin, when it was physically constrained from changing its conformation, and also had an n_H of only 1.0 (Chap. 6.3.1).

8.2.2 Structure of the Bacterial Phosphofructokinase

While the enzyme from *E. coli* was more thoroughly characterized for its kinetic features, the most informative structures were determined with the enzyme from *Bacillus stearothermophilus*.[212] These enzymes share some regulatory features, and the general structures for both the *E. coli* and *B. stearothermophilus* enzymes are quite similar. Because the separate structures have more detail, the *B. stearothermophilus* enzyme structure may be a useful model for understanding the kinetics with the *E. coli* enzyme. Figure 8.4 summarizes results obtained from different crystal structures of the enzyme, normally binding one ligand.

A number of features in this structural model are significant for understanding the kinetics and regulation of the bacterial enzyme. The ATP site is completely contained in one subunit, and in one domain in that subunit. This is consistent with no observable cooperativity with ATP; that is, the conformation of the ATP site remains constant. The enzyme clearly changes the conformation of the fructose-6-P site, as demonstrated by the cooperativity for this substrate (Fig. 8.2A). The conformational change that produces this results from the binding site requiring a loop on the adjacent subunit. It was described (Chap. 1.2.3.2) how any movement, such as rotation or bending, between subunits or domains in a single enzyme oligomer could result in forming or disrupting a binding site that occurs at the interface of the separately moving bodies. We see in Fig. 8.4 that if there is any separation between subunits 1 and 2, or any rotation between these subunits, then the site for the binding of fructose-6-P could be properly formed, under ideal conditions, or be disrupted under other conditions.

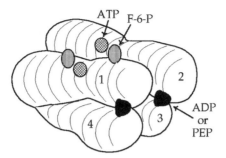

Fig. 8.4. Model for the tetrameric form of phosphofructokinase from *B. stearothermophilus*. The binding site for ATP is entirely on one subunit. Binding of fructose 6-P (F-6-P) requires part of one subunit plus a loop from the adjoining subunit. A regulatory site also occurs between subunits, where ADP produces activation, while phosphoenolpyruvate (PEP) produces inhibition. There are four sites for each ligand per tetramer

This feature becomes clear with the separate crystal structures shown in Fig. 8.5. There are two adjacent residues in a loop, Glu-161 and Arg-162 that are able to switch positions with each other as the loop flips over. With the free enzyme (Fig. 8.5A), Glu-161 is positioned very near to the site where the phosphate of fructose-6-P normally binds. The negative charge on Glu-161 makes binding of the negatively charged phosphate very unfavorable. This is the principal feature that defines the *T* conformation, and explains the normally poor binding of fructose-6-P to this conformation.

To obtain a structure of the *R* conformation, the enzyme was crystallized in the presence of ADP plus fructose-6-P, to simulate the quaternary complex of the enzyme with its two normal substrates (Fig. 8.5B). One can imagine a third phosphate on ADP,

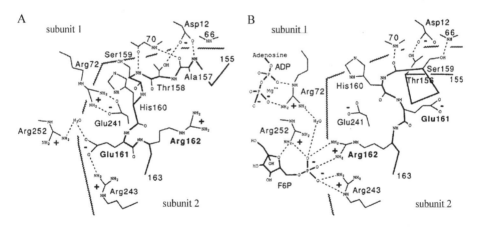

Fig. 8.5. Catalytic site of phosphofructokinase. (**A**) The free enzyme, in the *T* conformation. (**B**) The enzyme in the *R* conformation with ADP plus fructose-6-P at the catalytic site. Movement of a single loop removes the negatively charged glutamate-161, which is near the binding site for the phosphate of fructose-6-P, and replaces it with the positively charged arginine-162. The *bold jagged line* indicates the boundary between the two subunits (adapted by permission from Macmillan Publishers Ltd: Nature 343, 140–145, copyright 1990)

and visually position it so as to be near the carbon-1 position of the ribose ring of fructose-6-P. This would be the position of the gamma phosphate just prior to its transfer in the normal reaction. In the R conformation, as the allosteric loop flips to the alternate position, Glu-161 is moved away and Arg-162 is now positioned at the same location (Fig. 8.5B), and with its positive charge it can now interact favorably with the phosphate group on fructose-6-P. This leads to very favorable binding, and explains the stronger K_m of the R conformation.

These two structures are the two important species in the protein ensemble for phosphofructokinase. Each structure may be separately stabilized by ligands. As diagrammed in Fig. 8.4, the regulatory site on the bacterial enzyme is positioned at the end of each subunit, and requires part of the immediately closest subunit to form the complete site. The regulatory site is able to bind both a positive and a negative effector. To produce opposing kinetic results, for two different ligands at the same site, is possible if the ligands can either stabilize a more open form of the site, or a form that is more closed. There are separate crystal structures with the inhibitor phosphoenolpyruvate (PEP), and with the activator ADP. With the inhibitor PEP, the two domains (on separate subunits) that form the regulatory site rotate towards each other, as they are stabilized by the smaller PEP. By a direct physical linkage, this rotation leads to movement of the loop near the fructose-6-P site, to position Glu-161 so that binding of fructose-6-P is very unfavorable (equals inhibition).

Because ADP is sufficiently larger than PEP, binding of ADP stabilizes a different orientation of the two domains, as they separate a small distance from each other. This movement is also communicated to the loop at the catalytic site, such that it now favors the alternate position, in which Arg-162 is in the position to interact with the phosphate of fructose-6-P. This is a simple and elegant architectural solution to produce a protein that is still of modest size (compare to Fig. 1.5), and yet it has allosteric regulation controlled by an activator and an inhibitor.

8.2.3 Mutation to Reverse Regulation of Bacterial Phosphofructokinase

Because the crystal structure for the bacterial enzyme demonstrated how two opposing effectors could produce their regulation by binding at the same site, this suggested the possibility that a simple mutation at one of the key residues of this site might reverse the effects of the two regulators, so that ADP might be inhibitory, or PEP might become an activator. All six of the residues that participate in effector binding were mutated by Lau and Fersht,[213] and these mutants had fairly normal activity, as can be seen by comparing Fig. 8.6 with Fig. 8.2A. Some mutants also demonstrated a reversal in the regulatory response.

An example of one mutant (E187A) is shown in Fig. 8.6. The control curve has the normal positive cooperativity of the enzyme for the substrate fructose-6-P. When PEP was added, it clearly changed the conformation of the enzyme so that cooperativity is no longer evident. This would be expected if the enzyme is now largely in the R conformation.

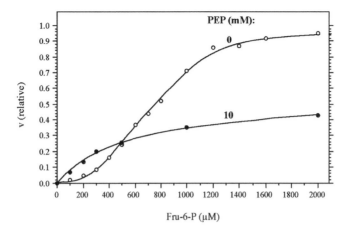

Fig. 8.6. Converting PEP into an activator with the E187A mutant of the *E. coli* phosphofructokinase. Data from Lau and Fersht[213]

Although the total activity of this mutant bound with PEP was only about one-half of the free enzyme, it is clear that at low substrate concentrations PEP caused the enzyme to have higher activity. This shows that at substrate concentrations that are too low to promote a significant shift from the *T* to the *R* conformation for the free enzyme, the addition of PEP has accomplished this change. Additional studies with this mutant demonstrated that GDP had now become an inhibitor.

8.2.4 Regulation of the Mammalian Phosphofructokinase

The mammalian enzyme has a subunit size more than twice as large as the bacterial enzyme (Table 8.1), and when the gene from rabbit was sequenced it became evident that the N-terminal half of the sequence had very high homology to the C-terminal half, and together they aligned very well with the sequence for the bacterial enzyme.[21] This showed that gene duplication had occurred, and that the two genes were contiguous and expressed as a single mRNA, and translated as a single large protein in mammals.

Such a duplication will then produce an enzyme with subunits that have twice as many binding sites. No crystal structures have been obtained for this larger phosphofructo-kinase, but a possible arrangement of the domains and the binding sites is illustrated in Fig. 8.7. The initial duplication event should have produced subunits with two catalytic sites on each, now larger, subunit. Since the second catalytic site was not essential, modest mutations in that site caused these sites to have only a regulatory function. It is important to assume that the regulatory linkages demonstrated for the bacterial enzyme (Figs. 8.4 and 8.5) have been retained in the mammalian subunits. In that sense, a single subunit of the mammalian enzyme is comparable to a dimer of the bacterial enzyme.

If we now examine Fig. 8.7, we see that the catalytic site has a similar arrangement as in the bacterial enzyme (Fig. 8.4). The ATP site is entirely on one domain, and

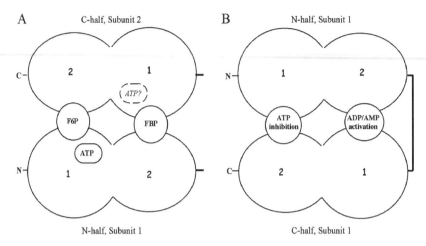

Fig. 8.7. A proposed model for the mammalian phosphofructokinase. Gene duplication plus fusion have resulted in a protein with twice as many binding sites as the ancestral bacterial enzyme. It is not clear if the ATP site on domain 1 of subunit 2 still functions. Adapted from Poorman et al.[21]

adjacent is the site for fructose-6-P, which again requires interaction with the domain on the neighboring subunit, shown on the left (Fig. 8.7A). In that same figure, on the right we see the two subsites of what had been a second catalytic site. The fructose-6-P site has been altered by mutation, and now binds the important activator fructose-2,6-P$_2$ (FBP). The ATP subsite may still bind ATP, but the consequence of ATP binding here is not clear. It may contribute to inhibition by ATP.

The new protein also has a duplication of the original regulatory site (Fig. 8.7B). One of these two subsites continues to bind AMP or ADP, where their binding promotes the *R* conformation and activation. The second subsite now binds ATP tighter, and this has become an inhibitory site for ATP. Therefore, the redundant catalytic site has been exploited to generate a novel regulatory site for a new activator, FBP. And the redundant regulatory site has become altered to bind ATP as the inhibitor.

8.2.5 How Many Binding Sites for ATP?

The bacterial enzyme appears to bind ATP only at the catalytic site. With no detailed crystal structures for the mammalian enzyme, how can we decide on the number of sites that actually function in binding any ligand? The best approach is to use pure enzyme preparations for binding studies with one single specific ligand. This has been done for the mammalian phosphofructokinase (Fig. 8.8). Binding studies were done with the native enzyme, resulting in a biphasic binding curve. It is clear, however, that the two data segments represent binding at three sites. The experiment was performed with 30 μM pure enzyme. If the enzyme had only one site for ATP, then there could only be a maximum of 30 μM ATP bound. We see that the data for the first binding segment with native enzyme extrapolates to 60 μM. This suggests that two sites with affinities that are strong enough for the experimental concentrations of ATP have become filled. A third

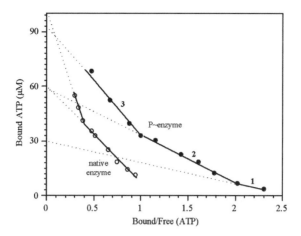

Fig. 8.8. Three separate binding sites for ATP on mammalian phosphofructokinase. The experiment was done with 30 μM enzyme, and in this Eadie–Hofstee plot an extrapolation is shown for each data segment to the ordinate, to show the total ATP bound as each site becomes filled. Data from Kitajima et al.[214]

binding segment is distinct, since it represents a poorer affinity for ATP, and when it is filled, we see that a little more than 90 μM ATP has been bound. This strongly implies that at least three distinct ATP binding sites exist.

Stronger support for three sites comes from the second experiment in this figure, using the phosphorylated form of the enzyme. Phosphorylation of phosphofructokinase leads to a conformational change, and increased affinity for ATP. This feature is directly evident since the binding curve for the P-enzyme is shifted to the right, with more ATP bound at any concentration of ATP. It is now possible to see that there are three data segments, for three separate binding sites. Although only two data points define the first binding segment, this extrapolates to 30 μM bound ATP, and indicates that one binding site has been filled. The second binding segment extrapolates to 60 μM ATP bound, and demonstrates that there must be at least two sites for the 30 μM concentration of enzyme subunits. The final binding segment extrapolates to ≥90 μM, and demonstrates that at least three distinct binding sites exist.

The value of such an experiment is that, in the absence of a protein structure, it may define a lower limit for the number of distinct ligand binding sites. We can assign the binding segments to specific sites on the protein. The site with the highest affinity (segment 1) must represent the catalytic site. If the inhibitory site had the highest affinity, the enzyme would not be very active at most ATP concentrations. The second segment shows weaker binding, and we have additional evidence that ATP becomes an inhibitor at higher concentrations (K_i = 1 mM). The third binding segment represents binding at very high concentrations of ATP. This very weak binding site is suggested to be the activator site (Fig. 8.7B),[214] although it may also be the remnant ATP site near the new FBP site (Fig. 8.7A). Binding at this third site is too weak, and this would not occur in vivo. It is, however, possible to demonstrate this with appropriate concentrations in vitro.

8.3 Key Regulators of Phosphofructokinase

8.3.1 Effectors for the Bacterial Phosphofructokinase

Since the bacterial enzyme has very little activity in the absence of an activator, ADP is normally used for kinetic studies (Fig. 8.2). To assist in understanding the relevance of the actual values for the various affinities, the K_d values are compared with the normal cellular concentrations of those compounds (Table 8.2). Although there will be some variation for the concentrations of these ligands in different bacteria, or in the different tissues of the same mammal, a single data set of concentration values is shown. For both the substrate, fructose-6-P, and the activator ADP, the normal concentration is similar to the K_d value. This suggests that in vivo the enzyme would be functioning near half of its maximum activity. This seems very appropriate for an important metabolic pathway.

By comparison, the concentration of PEP is very low compared to the K_i for this compound. This is also appropriate, since inhibition should not occur until this key metabolite is sufficiently abundant. PEP concentrations would normally increase under gluconeogenic conditions, and therefore signal that alternate fuel sources were available and being utilized. The important feature from these data is that the enzyme's affinities have evolved to be appropriate for the concentrations of the different ligands that are involved in regulating its activity, as was described in Chap. 2.2.2.

8.3.2 Effectors for the Mammalian Phosphofructokinase

The mammalian enzyme has been studied extensively by Uyeda and coworkers.[206, 214, 215] We will use the suggested architecture (Fig. 8.6), to compare the binding of the various regulators with their cellular concentrations (Table 8.3). The results with the substrate fructose-6-P and AMP, a nucleotide activator comparable to ADP, show that the bacterial features have largely been retained. Under normal conditions, the enzyme will be at least 50% bound to AMP. However, AMP itself is not sufficient to fully activate the enzyme. This is evident in the experiment shown in Fig. 8.9. These studies were done with the enzyme from rat muscle,[206] and demonstrate that with AMP at concentrations equal to or greater than in vivo, the enzyme achieves less than 20% of its full activity.

Table 8.2. Regulators for the bacterial phosphofructokinase[a]

Compounds	Binding site	Affinity[b]	Cell. concentration[c]
Activators			
ADP	regulatory site	$K_R^{ADP} = 25\,\mu M$	$<100\,\mu M$
Fructose-6-P	catalytic site	$K_C^{F6P} = 12.5\,\mu M$	$4\,\mu M$
Inhibitors			
PEP	regulatory site	$K_R^{PEP} = 750\,\mu M$	$15\,\mu M$

[a] From Blangy et al.[203]
[b] Abbreviations: C, catalytic site; R, regulatory site
[c] Values are from mammalian cells

The mammalian phosphofructokinase responds to several inhibitors (Table 8.3). ATP and citrate are the most important, and an increased abundance of these signifies that either the ATP levels are adequate for cellular needs, or that other fuels are available and being converted to citric acid cycle intermediates. Note that the K_d values for these inhibitors are appropriate for their normal cellular concentrations.[§]

It is unclear if the pH effect (Table 8.3) is mediated by residues that immediately influence the catalytic site, or whether this is the more general result of a protein being sensitive to alterations in pH, as this may influence its ability to maintain a normal conformation. The enzyme has an optimum activity at pH 8, and the activity declines 90% as the pH is lowered to 6.8. This results in limiting the continued production of lactic acid when actively working muscles become acidic.

8.3.3 Effectors of the Yeast Phosphofructokinase

With the ATP-PFK, P_i may bind at any site where the phosphates of ATP normally binds, and thereby prevent ATP from having an effect. This enables the enzyme from yeast to use inorganic phosphate to block inhibition by ATP. This is physiologically significant, since higher concentrations of P_i occur when ATP is being depleted. Under normal assay conditions with 1.0 mM ATP, 10 mM P_i produces a 20-fold increase in activity.[216] If 1 mM AMP is also included, the effect due to P_i increases activity 100-fold.

8.3.4 Regulation by Phosphorylation

Covalent modification by phosphorylation is the most widely used mechanism for a transient change in an enzyme structure. The overall effect of phosphorylation is to make the enzyme in liver less active. This permits glucose derived from glycogen, or from gluconeogenesis to be exported to the blood and used by other tissues. To cause this, phosphorylation is initiated in response to extracellular hormones. For liver, epinephrine and glucagon are the major hormones that initiate phosphorylation of the enzyme, in a fashion depicted earlier (Chap. 4.1.3). Phosphorylation results in a conformational change that affects almost all of the ligand binding sites on the enzyme (Table 8.4).

The overall effect of phosphorylation is that the substrate and activators bind less well, and that inhibitors bind better. The absolute changes in affinity for the five ligands in this table are all threefold or less. By themselves, such changes in affinity would have a corresponding effect on the enzyme's activity. Since they occur at different binding sites, their cumulative effect has the potential to be synergistic, though this remains unquantified.

[§]It is the free concentration of the metabolites that is significant. For many metabolites the free concentration may be ≥50% of the total. But, ATP is used by so many enzymes that the total ATP concentration of 3–5 mM[31] would be very misleading.

Table 8.4. Changes in affinity with phosphorylation of phosphofructokinase[a]

Enzyme	$S_{0.5}^{F6P}$ (μM)	K_A^{AMP} (μM)	K_A^{FBP} (nM)	K_I^{ATP} (mM)	$K_I^{citrate}$ (μM)
Native PFK	9	8	6	1.6	290
P~PFK	12.5	26	10	0.8	75

[a]From Sakakibara and Uyeda[215]

8.4 Isozymes

There are at least three isozymes in mammals, and data for the human forms are listed in Table 8.5. A mixed nomenclature is currently in use. Many papers refer to the A, B, C isozymes, but others prefer the use of L (liver) and M (muscle), since these letters are intended to identify the major tissue for that isozyme. This is not absolutely correct, since platelets and other blood cells have a greater proportion of the L isozyme than liver (Table 8.5). Western blots with antibodies for the individual isozymes have shown that in most tissues all three are expressed, and helped to define the absolute abundance of each species.[217]

The data (Table 8.5) show that the muscle isozyme has the best affinity for fructose-6-P and shows the poorest sensitivity to inhibition by ATP. These values appear to be consistent with our expectation that both skeletal muscle and heart muscle must be capable of the greatest energy output from glucose, and the muscle isozyme can achieve a greater level of activity for the same concentrations of the two substrates. The L isozyme has intermediate affinity, and the C isozyme has the poorest affinity for fructose-6-P and the highest sensitivity to inhibition by ATP. Note that the C isozyme, which has been associated with the brain, where it was first discovered, is not the major isozyme in the brain. It is more prevalent in blood cells that do not normally experience significant changes in their need for energy. The brain has more than 50% of its phosphofructokinase of the M type, and therefore has kinetics that would be more similar to those for muscle. This is consistent with the pattern that we observed for the isozymes of glycogen phosphorylase (Chap. 7.3.1).

Table 8.5. Isozymes of human phosphofructokinase

Isozyme	Location[a]	Tissue: oligomers[b]	$S_{0.5}^{F6P}$ (mM)	K_I^{ATP} (μM)
M (A)	Muscle, tissues	Muscle: M_4	2.7	50
L (B)	Liver, tissues	Liver: L_2MC		
		Placenta: L_4	3.5	35
C	Brain, platelets	Platelets: C_2LM	5	25
		Brain: M_2CL		

[a]The principal tissue(s) are listed, though each isozyme may occur at lower quantities in most or all tissues. Data from Dunaway et al.[217]

[b]Where multiple isozymes occur, this denotes their average abundance in tetramers, and more varied compositions of the tetramer are possible

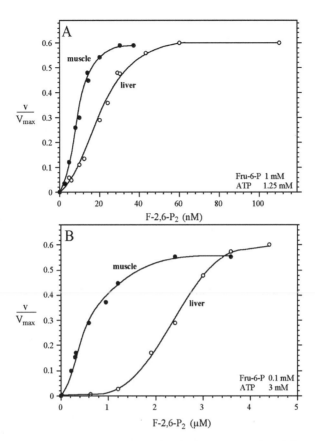

Fig. 8.11. Varied affinity for the principal activator, fructose-2,6-P$_2$ by muscle and liver isozymes. Not the changes in both substrates from (**A**) to (**B**) as ATP becomes an inhibitor. Data from Uyeda et al.[20]

8.4.1 Binding of Fructose-2, 6-P$_2$ by Mammalian Isozymes

Isozymes vary not only in their affinities for the principal ligands, but also in how cooperative they are in binding these ligands. This is illustrated in the binding curves for the key activator, fructose-2,6-P$_2$ with isozyme preparations from muscle and liver (Fig. 8.11). Although the curves were identified by these isozyme types in the original publication,[206] we are aware that the liver preparation would have contained a mixture of isozymes (Table 8.5). We see that the liver enzymes have poorer affinity for this key activator, and also bind it with strong positive cooperativity. By comparison, not only the muscle isozyme has better affinity, as also shown in Table 8.5, but also the normal conformation of the muscle isozyme has almost the optimum affinity for this activator, since cooperativity is more limited. That is, the binding curve is close to showing constant affinity for FBP.

The two experiments (Fig. 8.11) were intended to observe the affinity toward the principal activator, FBP, by the two isozymes under conditions where fructose-6-P was more abundant with ATP at a lower concentration (Fig. 8.11A), and then with fructose-6-P at a low concentration and ATP high enough to be inhibitory (Fig. 8.11B). The authors estimated that the conditions of the second experiment were closer to cellular conditions in vivo. Note the change in the units for the two panels. It is then interesting that the affinity of both isozymes is much greater when fructose-6-P is saturating, with ATP at 1.25 mM (Fig. 8.11A). Presumably this reflects that ATP has initiated little binding at its effector site. By comparison, when fructose-6-P is merely at a high concentration and ATP is now at a concentration high enough to be a good inhibitor, the affinity of both isozymes for FBP has weakened nearly 50-fold.

This then tells us that binding at the ATP inhibitor site is dominant over any results due to binding of FBP at its regulatory site. Such a linkage of control was also seen with glycogen phosphorylase, where binding at the inhibitory site was normally weaker, but at high effector concentrations, inhibition was the final state (Fig. 7.4). Again, this is physiologically sensible. Whenever ATP concentrations are sufficiently high, further ATP production by glycolysis is no longer desirable. Therefore binding of ATP at the inhibitory site stabilizes the T conformation, no matter what other ligands are interacting with the enzyme. Citrate is also an important inhibitor, and it is very likely that citrate binds to the ATP inhibitory site, though this has not been verified. But this is a very reasonable assumption, since the three negative charges on citrate would help it bind at the ATP inhibitory site and produce the same results.

Overall, we see the benefit of isozymes since for many tissues a mixture of the different species is evident, and this permits a greater range and a more gradual change in the sensitivity for any of the substrates or regulatory effectors. While the 2-log rule remains a useful criterion for the binding of any single protein to a ligand, we see that Nature continues to develop enzyme systems that are not limited by this constraint. In addition to having some of the enzymes become allosteric, and thereby have a greater or narrower ligand concentration range, the evolution of isozymes that have become slightly modified extends this strategy to have more complex variations in the response to any concentration of a given ligand.

8.5 Defining a Model for Cooperativity

For the enzyme from *E. coli*, Monod and colleagues used the equations from their model for a two-state system to fit curves to the kinetic data, with very good results (Fig. 8.2). As judged by eye these figures suggest that the two-state model appears quite satisfactory to describe the allosteric features of this enzyme. However, certain experiments by other investigators require more than two conformations to achieve a full description.

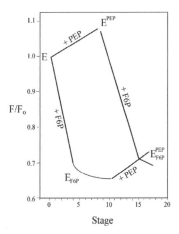

Fig. 8.12. Change in intrinsic fluorescence with ligand binding by the *E. coli* phosphofructokinase. Adapted from Johnson and Reinhart[218]

8.5.1 Fluorescence Changes with Different Ligands

The enzyme from *E. coli* has only one tryptophan, so that changes in the fluorescence of this residue as a function of conformational change can provide a direct measurement of possible states that the protein may attain. In these studies the fluorescence of the free enzyme (E) has a normalized value of 1.0. Fluorescence may increase or decrease as the changes in the enzyme's structure alter the immediate environment of the tryptophan being observed. With continued addition of the substrate fructose-6-P, the fluorescence becomes quenched, as shown in Fig. 8.12, until a new end point is achieved at E_{F6P}.

If PEP is added to the free enzyme, fluorescence increases about 7% as the enzyme becomes the E^{PEP} complex. Addition of F6P to the latter complex then initiates quenching as the system reaches the ternary complex E^{PEP}_{F6P}. In a comparable fashion, if the enzyme is initially titrated with F6P, the subsequent addition of PEP will increase the fluorescence, and reach a very similar fluorescence end point as the complete ternary structure is again formed.

Therefore, with only two of the possible ligands interacting with the enzyme, the fluorescence changes demonstrate that at least four different conformations must exist to explain these results. If then additional ligands were permitted in this mixture (ATP, FBP) more complex enzyme species would be produced and additional conformational states are very possible.

9

RIBONUCLEOTIDE REDUCTASE

Summary

Ribonucleotide reductase catalyzes the essential step in the formation of all four deoxynucleotides. Although the earliest form of the enzyme emerged in a reducing environment, additional forms evolved to function in an aerobic state. While the different classes of this enzyme have different cofactors, and different mechanisms for generating the free radical necessary for the reaction, they have retained many of the key regulatory features. Most of the microbial enzymes have one regulatory site, while the enzyme from *E. coli* and most eukaryotes normally has two. Regulation at these sites by nucleotide effectors permits the enzyme to alter its selection for any of the four ribonucleotide substrates.

9.1 The Diversity of Ribonucleotide Reductases

Ribonucleotide reductase activity replaces the 2'-hydroxyl on the ribose of ribo-nucleotides with a proton to form deoxynucleotides. The enzyme for this reaction is essential in the overwhelming majority of organisms that use DNA for storage of genomic information. There are four distinct classes of this enzyme based on unique requirements for the metal cofactor, the mechanism for reducing the reactive cysteines, and the size and structure of the subunits (Table 9.1). It is therefore remarkable that such variations in a catalytic process can be reconciled with a model for divergent evolution of these enzymes from a common ancestor.[219, 220]

Protein structures have recently been determined for at least one enzyme from each subclass (described in Ref. 220). The class Ia enzymes remain the best characterized for their kinetic and regulatory properties, and will therefore be used for most of our interpretations. The type Ia enzymes are $\alpha_2\beta_2$ tetramers, where the α_2 dimer contains the catalytic site and the regulatory sites, while the beta subunits contain the elements for producing a free radical. A scheme for the α_2 structure is shown (Fig. 9.1).

Table 9.1. Types of ribonucleotide reductases[a]

	Class Ia	Class Ib	Class II	Class III
Oxygen need	Aerobic	Aerobic	Aerobic/anaerobic	Anaerobic
Structure	$\alpha_2\beta_2$	$\alpha_2\beta_2$	α or α_2	$\alpha_2 + \beta_2$
Substrate	NDP[b]	NDP	NDP/NTP	NTP
Electron donor	Thioredoxin Glutaredoxin	Glutaredoxin	Thioredoxin	Formate
Radical	Tyrosine on β	Tyrosine on β	Ado-cobalamin	Glycine on α
Metal site	Fe–O–Fe	Fe–O–Fe Mn–O–Mn	Co	Fe–S
Allosteric sites/α subunit	2	1	1	1

[a]Varied names have been used for the different subunits: α represents B1 or R1 in other publications, while β represents B2 or R2

[b]NDP, nucleoside diphosphate; NTP, nucleoside triphosphate

Three separate binding sites are evident on each subunit. The catalytic site (C) at which the actual chemistry occurs is located between domains, a structural feature that facilitates conformational changes, and which we have seen as a common feature in other allosteric enzymes such as glycogen phosphorylase and phosphofructokinase.

There are also two regulatory sites. The main regulatory site is the specificity site (S). This site is found on all the different enzymes, and is positioned near or at the interface between the two alpha subunits, so that binding of a ligand influences the shape of a loop from the adjoining subunit, as diagrammed in Figs. 9.1 and 9.2. Loop 2 extends between the S and C sites, so that there is physical contact that is influenced by ligands binding at each site. It is the flexibility of this loop that permits the C site to vary in its ability to bind each of the possible NDP substrates and enables different combinations of nucleotides to simultaneously occupy the S and the C site (Fig. 9.2), and make it possible for each of the effectors to specify which substrate will be preferred.

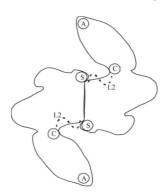

Fig. 9.1. A diagram for the α_2 dimer of the type Ia enzyme (from *E. coli*), based on the crystal structure.[221] Since the human enzyme belongs to this subclass it will presumably have a similar structure. The subunit has three domains, with the catalytic site (C) between two domains. Regulation is by effectors binding at the activity site (A) or at the specificity site (S). Binding at the specificity site influences the choice of substrate at the catalytic site via loop 2 (L2) on the adjacent subunit

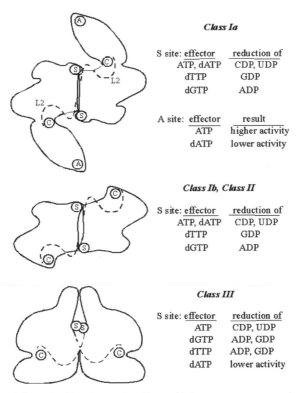

Class Ia

S site: effector	reduction of
ATP, dATP	CDP, UDP
dTTP	GDP
dGTP	ADP

A site: effector	result
ATP	higher activity
dATP	lower activity

Class Ib, Class II

S site: effector	reduction of
ATP, dATP	CDP, UDP
dTTP	GDP
dGTP	ADP

Class III

S site: effector	reduction of
ATP	CDP, UDP
dGTP	ADP, GDP
dTTP	ADP, GDP
dATP	lower activity

Fig. 9.2. Regulation of ribonucleotide reductase by effectors binding at one or two regulatory sites. For each class the alpha subunits are shown. The *dashed line* represents Loop 2, by which regulation for particular nucleotide substrates is achieved. For the class III enzyme the S site (for the subunit at right) is toward the back, while the S site (for the subunit at left) is toward the front[222]

The ability to balance deoxynucleotides is essential for all cells or organisms as they prepare for DNA duplication. The fidelity of duplicating DNA becomes challenged by serious imbalances in the concentrations of the four deoxynucleotides. When the polymerase is at a position for adding a nucleotide, if the correct nucleotide is at a low concentration while any other nucleotide is at a much higher concentration, then misincorporation will occur at a much greater frequency. Although polymerases are known for their proofreading functions, as misincorporation rates increase, the overall mutation rate will still increase. Maintaining balanced deoxynucleotide pools is therefore useful to prevent increased mutation rates. This was necessary even with the earliest organisms, since all existing ribonucleotide reductases have this regulatory feature.

Type Ia enzymes also have an activity site (A) that controls the overall rate for reductase activity with any of the four possible substrates. This site is located at one end of the ATP domain. This is the major binding site for ATP, whose effect is to increase the enzyme's rate. In contrast, binding of dATP at the A site generally has an inhibitory effect. These opposite effects are very logical, as shown by the values in Table 9.2. An

Table 9.2. Binding constants for nucleotide effectors[a]

Nucleotide	Concentration (μM) in vivo[b]	K_d (μM) S site	K_d (μM) A site
ATP	≤200	>10[c]	10
dATP	3.2	0.03	0.5
dGTP	1.5	0.08	Not defined
dTTP	5.4	0.3	≥200 μM[c]

[a]Binding data from Brown and Reichard[226]

[b]These values are for cultured cells that are well nourished and mitotic. Although these are average values for such cells, they will be high compared to normal or nonmitotic cells[31]

[c]Estimated from original data

abundance of ATP indicates that the bacterial cell is well nourished, favoring proliferation by cell division. In contrast, an abundance of dATP indicates that deoxynucleotides are at a high enough concentration to enable mitosis and cell division, so that further deoxynucleotide production is no longer needed. Kinetic studies clearly define that the S site distinguishes the difference at carbon 2 of the ribose between ATP and dATP.

Structures for the class Ib, class II, and class III enzymes show some variation. The class Ib/II enzyme has a smaller alpha subunit, and as diagrammed (Fig. 9.2) the A site is missing because part of the protein sequence is missing.* The catalytic site (C) occurs at a position where an extra domain could be positioned, comparable to the structure for the class I enzyme. However, as suggested by the diagram, the structure of the class Ib/II enzyme shares features with the class Ia enzyme. Both the catalytic site and the specificity site are positioned so that regulation by effectors at the S site alters the position of the L2 loop on the adjacent subunit. This may be the main structural feature that explains why binding of effectors at the S site has the same selective effect on the substrate for both class Ia and class Ib/II enzymes.

The class III enzyme has a somewhat different structure, in terms of the overall shape, yet again the C site is located between domains, and it is connected by a dynamic loop to the S site. The S site itself is at the interface of the two subunits, though there is still one such site per subunit. We can also note that the specificity for the substrate determined by the effector at the S site is less exact, since dGTP and dTTP produce the same result. However, while the results of effectors at the S site are more exact for class Ia and class Ib/II enzymes, their effects are consistent with the broader regulatory pattern for class III. This is one of the features in support of the divergent evolution model, with class III being the earliest.[223]

---

*It is thought that both class Ia and class Ib/II enzymes are descended from class III, whose subunit has no A site. It is not established if class Ia enzymes came next and evolved the A site, to then be lost with class Ib/II, or if class Ib/II came second, so that the A site evolved only in class Ia enzymes.

9.1.1 Evolution of Ribonucleotide Reductases

The mechanism involves the generation of an organic free radical that is then transferred to a cysteine at the active site. The actual formation of the radical occurs by different cofactors in each of the subclasses (Table 9.1). The type III enzyme, considered most likely to be the ancestral enzyme,[224] had to be anaerobic, and used formate to produce a radical on a glycine residue by steps also seen with pyruvate formate lyase. Since this lyase is considered one of the oldest enzymes needed in the simplest and earliest cell, it has been proposed that the ancestral ribonucleotide reductase evolved from such a lyase.[224]

As oxygen became important in the environment (see Chap. 6.1.1) class III enzymes were sensitive to damage by oxygen,[†] and class II enzymes emerged and used adenosyl–cobalamin for generating the free radical, and thioredoxin as the source for the electron. Further changes led to the class I enzymes and use of Fe–O–Fe at the metal site, and a tyrosine radical. While the source of the free radical has changed, in each case this still leads to the radical appearing on a cysteine as part of the catalytic mechanism.[225]

9.2 Binding of Nucleotides to Ribonucleotide Reductase

Since the type Ia enzyme has three binding sites for either ribonucleotides or deoxynucleotides, activity experiments might be difficult to interpret in terms of which molecules are binding at the different sites, since at least two different nucleotides must be used: one as the substrate, and a second nucleotide as an effector, to ensure enough activity for reliable measurements. If binding experiments are done with a single nucleotide at one time, the interpretation becomes clearer.

9.2.1 Binding of dATP

The type Ia enzyme from *E. coli* is the best characterized for its regulation and will be the main focus for analyzing how regulation is achieved. Since the human enzyme belongs to the same subclass, it presumably shares many or all of these features. The earliest studies with purified enzyme initially established that both ATP and dATP appeared to bind to at least two sites on the alpha subunit, with opposing results. ATP activates the enzyme towards all four substrates. dATP at concentrations below 1 µM also activates the enzyme, but becomes an inhibitor at concentrations above this value.[225]

To elucidate these different effects Peter Reichard and his group performed a combination of binding experiments and inhibition experiments to help define a model for the structure and function of the enzyme that could explain their results. As shown by the binding of dATP to the type Ia alpha dimer (Fig. 9.3), two binding sites per subunit are evident for the concentrations of the nucleotide used in the experiment. Total binding did not change when the binding experiment was done with the complete $\alpha_2\beta_2$ tetramer, demonstrating that the beta subunit has no binding site for this nucleotide. The addition

---

[†] In the presence of oxygen the glycyl radical results in cleavage of the peptide backbone.[225]

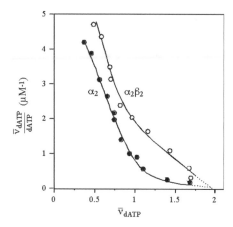

Fig. 9.3. Binding of dATP to the type Ia α_2 dimer, or $\alpha_2\beta_2$ tetramer at 4.4 µM. Data from Brown and Reichard[226]

of the beta subunit did change the affinity of the alpha subunits for the nucleotide ligand. The binding curve is biphasic, evidence that there is a significant difference in the K_d for the two sites. While these two sites were initially named h (for high affinity) and l (for low affinity), we will employ the most recent nomenclature for the enzyme (shown in Figs. 9.1 and 9.2), since the high or low site can vary for different nucleotides.

As shown in Table 9.2, ATP binds better at the A site while dATP and other deoxynucleotides bind better at the S site. Therefore, the original designations of high and low affinity sites are reversed for these two effectors. Also shown in Table 9.2 are average cellular concentrations for these nucleotides, though these values are mostly from mammalian cells.[31] Using these cellular concentrations we see that ATP will normally saturate the A site, but only partly occupy the S site. Using the binding data shown in Fig. 9.4 and the concentrations in Table 9.2, we see that dATP will generally occupy the S site completely. The A site has an appropriately lower affinity, so that it is more sensitive under cellular conditions to the normal changes in the concentration of dATP.

9.2.2 Binding of ATP

Binding of dATP shows negative cooperativity (Fig. 9.3) since the second site has much poorer binding. In contrast ATP shows classic positive cooperativity in binding to the α_2 dimer (Fig. 9.4). The K_d of 10 µM was determined for binding of ATP to the first site (the A site).[226] How do we distinguish if the cooperativity is for improved binding at the second A site in the enzyme dimer, or at the different S sites in the dimer? Here we see a benefit of plotting such binding data in the Eadie–Hofstee or Scatchard format, as shown in Fig. 9.4A. Cooperativity is clearly evident before one site per subunit is fully occupied with ATP, showing that this cooperativity is likely to be between the two A sites on subunits in the dimer. There is then additional cooperativity, suggesting continued or additional conformational change, to promote binding of ATP at the S site. This

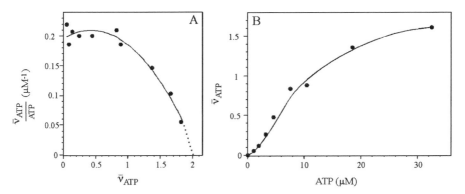

Fig. 9.4. Binding of ATP to the type Ia α_2 dimer. (**A**) Eadie–Hofstee plot; (**B**) linear plot. Data from Brown and Reichard[226]

interpretation is consistent with the data showing that three deoxynucleotides bind fairly tightly at the S site. By comparison, ATP has a poorer affinity at its own A site, and an even lower affinity at the S site, which presumably is designed for deoxynucleotides.

When ATP binding data are plotted in a direct linear plot we see the normal sigmoid curve indicating positive cooperativity (Fig. 9.4B). We also see that the $K_{0.5}$ is just above 10 μM, a value fairly similar to the binding constant for the A site (Table 9.2). This suggests that binding at the S site is not that much poorer, so that the conformational effect has significantly improved the binding of ATP at the S site. If we expect cellular free ATP concentrations to be at least 100 μM, then the A site will be mostly filled with ATP. While binding of ATP at the S site has been demonstrated in the absence of deoxynucleotides (Fig. 9.4), at normal cellular concentrations for all these nucleotides ATP is less likely to have significant binding at the S site.

The original work did not extend ATP to concentrations above 30 μM, and so we do not know if ATP could also bind, if poorly, at the catalytic site. However, since free cellular ATP concentrations are seldom above 200 μM, it is then unlikely that in vivo ATP could have any significant binding at the C site.

9.2.3 Binding of dTTP and dGTP

The regulatory nucleotide dTTP appears to bind to only one site on the alpha subunit (Fig. 9.5A), and dGTP also binds to a single site (Fig. 9.5B). There is a single data point in Fig. 9.5A to imply binding at a second site, but this experimental concentration of dTTP was not high enough to make this more definitive. For the binding of dTTP we see the curvature that defines negative cooperativity. This would result if the affinity at the A site (the second site) is sufficiently weaker (Table 9.2). However, negative cooperativity is already evident before a single site has been fully bound.

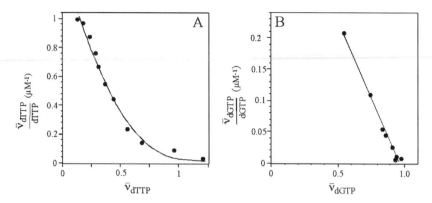

Fig. 9.5. Binding of dTTP (**A**) and dGTP (**B**) to the type Ia α_2 dimer. Data from Brown and Reichard[226]

The authors mentioned that the enzyme is unstable, and that great care was necessary to keep it at a normal activity.[226] It is therefore very possible that part of the change in affinity demonstrated in Fig. 9.5A is an example where negative cooperativity is an artifact produced by an enzyme sample that is partly denatured. We therefore see negative cooperativity for the full binding of the S site due to part of the enzyme having become unstable. We also see a suggestion of negative cooperativity for binding of dTTP at a second site where the affinity is poorer. See Chap. 5.3.2 for a description of such kinetics.

It is interesting that a similar experiment with dGTP gives the expected result for binding at a single site with constant affinity. If dATP and dTTP are able to bind at the A site, it is quite possible that dGTP also binds there. If the affinity of dGTP at the A site is sufficiently weaker, then binding at this second site would not be detected at the lower experimental concentrations of dGTP employed.

If we examine Fig. 9.5 we can also see that the affinities of dGTP and dTTP for the enzyme are not very different. The experiment with dTTP was performed over a greater concentration of the ligand, but we see that when the ratio of bound to free dTTP is about 0.2, the binding site is about 50% occupied. This is similar to the lowest concentration of ligand in the dGTP experiment at which bound/free equals 0.2, and the S site is also about 50% filled.

9.2.4 Competition Between Regulatory Nucleotides

In additional binding experiments we see how a second nucleotide, ATP, competes against a nucleotide known to bind at the S site. Figure 9.6 shows the binding of both ATP and dTTP as the concentration of total ATP increases. The initial dTTP concentration remained fixed at 5 μM for this experiment, which is adequate to nearly saturate the S site with dTTP ($K_d = 0.3 \, \mu M$). As the ATP concentration exceeds 10 μM, ATP itself begins to fill the A site, while the binding of dTTP is still very good, since each ligand now occupies one site on the enzyme. Experiments such as this helped to establish that ATP and the deoxynucleotides have their highest affinity for two different binding sites.

As the concentration of ATP reaches 200 μM, the A site is now nearly fully occupied by ATP, and dTTP binding is decreasing, as it is becoming modestly displaced at the S site by ATP binding there. Only when ATP is at 2 mM it is able to displace more than half of the dTTP, which is present at only 5 μM. These binding curves demonstrate the binding constants in Table 9.2.

Similar experiments show competition between ATP or dTTP vs. dATP (Fig. 9.7). dATP, which has the best affinity at the S site (Table 9.2) almost fills this site in the absence of the competing nucleotides. The two curves showing release of dATP as either dTTP or ATP is increased are similar, although dTTP is about 100-fold more effective because it binds that much better at the S site. A careful comparison of Figs. 9.6 and 9.7 shows that 2 mM ATP is sufficient to completely displace dATP, but only partly displace dTTP. These results are inconsistent with the average K_d values for dATP and dTTP (Table 9.2). That is, since dATP is assigned the better affinity at the S site, it should be the nucleotide that is more difficult to displace by ATP. Although these experiments were done by the same scientists, and with similar enzyme preparations, such disparities are likely the result of occasional difficulties with maintaining enzyme stability.

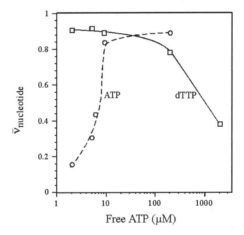

Fig. 9.6. Binding of ATP or dTTP to the type Ia α_2 dimer. Initial dTTP was at 5 μM. At the lowest ATP concentrations dTTP occupies one subunit site almost fully (S site), but ATP binds to a second site (A site). Only at ATP $\geq$ 100 μM does ATP begin to displace dTTP from the S site. Data from Brown and Reichard[226]

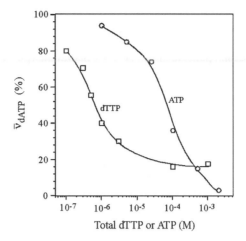

Fig. 9.7. Binding of dATP to the type Ia α_2 dimer, in the presence of competing dTTP or ATP. The initial free dATP concentration was 0.7 µM. Data from Brown and Reichard[226]

9.3 Regulation of Ribonucleotide Reductase from *E. Coli*

Having established the affinity and binding sites for the principal regulatory nucleotides (above) it is now possible to see how these effectors influence the activity of the enzyme with the substrate CDP. In most of the following figures two regulatory effectors will be examined in each experiment. The observed activity for the formation of dCDP is always below 0.4 s⁻¹, even in the presence of positive effectors. Although the enzyme was shown to be pure, the preparations for this enzyme in the late 1960s had not achieved optimum stability, and we see only a very low level of activity for this important metabolic enzyme.

9.3.1 Regulation by ATP and dATP

The combined effects of ATP and dATP are shown in Fig. 9.8. When dATP is absent, then ATP becomes an activator at concentrations ≥ 1 µM, the lowest concentration at which it begins to bind at the A site ($K_{A\,Site}^{ATP} = 10$ µM). For this curve the effect of ATP is almost maximum when the concentration of ATP is 100 µM, and has almost saturated the A site. As ATP is increased to 5 mM it would also saturate the S site, yet we see no significant increase in activity due to binding at a site that also increases the affinity for specific substrates. The value of 0.4 s⁻¹ represents the highest activity for this enzyme preparation, when both regulatory sites are occupied.

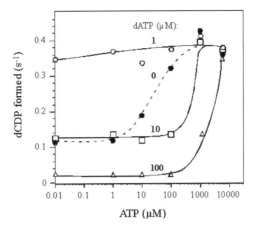

Fig. 9.8. Regulation of activity by ATP at fixed concentrations of dATP, and the type Ia α_2 dimer at 0.03 nM. ATP activates if dATP concentrations are >1 µM. Data from Brown and Reichard[227]

When dATP is present at 1 µM, the enzyme also has this optimum activity, and the presence of ATP has little influence. A 1 µM concentration of dATP is enough to fill the S site ($K_{\text{S Site}}^{\text{dATP}}$ = 0.03 µM), while also binding at least half of the A site ($K_{\text{A Site}}^{\text{dATP}}$ = 0.5 µM). Although ATP is considered to be an activator, the fact that its binding at the A site provides no increase in total enzyme activity is again evidence that for activating the enzyme the S site is more important, and if this site is near saturation, the enzyme is near its optimum activity and binding at the A site cannot improve this conformation for the active site.

An experiment similar to that in Fig. 9.8 was performed, but with dATP as the variable (Fig. 9.9). With ATP absent, dATP at concentrations below 1 µM activates the enzyme. For this experiment dATP would saturate the S site ($K_{\text{S Site}}^{\text{dATP}}$ = 0.03 µM), while occupying the A site about 50% ($K_{\text{A Site}}^{\text{dATP}}$ = 0.5 µM; Table 9.3). It is then the increasing occupancy by dATP of the A site that limits the total activity. This interpretation is supported by the two experiments in Fig. 9.9 where ATP is present at 1 or 10 µM. At these two concentrations ATP begins to modestly occupy the S site, while having a greater presence at the A site (Table 9.3). Since these latter two experimental curves reach their optimum with dATP at 0.4 µM, this must reflect the inhibition that becomes greater as dATP increases its occupancy at the A site. While ligand binding at both sites influences the conformation of the enzyme (as judged by the change in activity), occupancy by dATP at the A sites provides a limit to the highest activity.

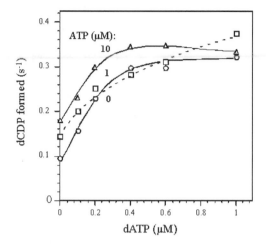

Fig. 9.9. Regulation of activity by dATP at fixed concentrations of ATP, and the type Ia α_2 dimer at 0.3 nM. At dATP $\leq$ 1 μM it activates for all concentrations of ATP. Data from Brown and Reichard[227]

9.3.2 dTTP at the S Site Promotes the Reduction of CDP

To define the actions of separate deoxynucleotides, Reichard and colleagues monitored the reduction of CDP in the presence of effector nucleotides that could bind at the S site or at the A site.[226, 227] By itself dTTP produces increased activation until an optimum is achieved with dTTP at 30 μM (Fig. 9.10). Increasing the concentration of dTTP up to 1 mM provides no additional activation, and appears to have a very slight inhibitory effect. The mid-point for the activation effect is at about 0.3 μM, the value for the affinity of dTTP at the S site (Table 9.2). Since saturation for the activation effect does not occur until dTTP increases 2 logs above the K_d for the S site, this kinetic experiment shows negative cooperativity for the effect by dTTP. This result is then similar to the negative cooperativity observed with the binding of dTTP (Fig. 9.5A).

9.3.3 dTTP at the A Site Inhibits the Reduction of CDP

When ATP is present at 0.1 or 1 mM, the ATP itself provides maximum activation, so that increasing dTTP concentrations have no effect until they reach greater than 50 μM. This very high concentration implies that the S site should be completely filled by dTTP, and

Table 9.3. Saturation of the S and A sites for the experiment in Fig. 9.9

Nucleotide	Concentration in Fig. 9.9 (μM)	Occupancy S site (%)	Occupancy A site (%)
ATP[a]	0	0	0
	1	$\leq$2	~10
	10	$\leq$10	50
dATP	0.5	~100	50

[a]The ATP affinity at the S site is not well defined, but appears to be at least fivefold poorer than at the A site

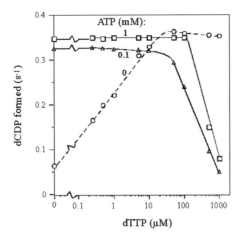

Fig. 9.10. Regulation of dCDP synthesis by dTTP at fixed concentrations of ATP, and the type Ia α_2 dimer at 0.3 nM. In the absence of ATP dTTP activates the production of dCDP. At higher ATP concentrations dTTP inhibits when it is >100 μM. Data from Brown and Reichard[227]

therefore suggests that dTTP can also bind at the A site, where it has an inhibitory effect similar to dATP. For these two experiments showing inhibition by dTTP (Fig. 9.10) the concentrations of ATP are such that ATP should almost fill (ATP = 0.1 mM) or completely fill (ATP = 1 mM) the A site. The increasing concentrations of dTTP are then able to bind at the A site, and displace ATP, thereby stabilizing the same inhibitory conformation produced by dATP. While the data are not sufficient to provide an accurate K_i, this value for the A site should be in the range of 200 μM.

Figures 9.11 and 9.12 show results with dTTP, when the second nucleotide effector is either dATP or ATP. The enzyme has a residual level of activity, about 0.1 s^{-1}, when neither effector is present. If only dTTP is added, then the presence of 1 μM dTTP doubles the activity (Fig. 9.11). This concentration of dTTP would provide greater than 50% occupancy at the S site ($K_{\text{S Site}}^{\text{dTTP}}$ = 0.3 μM). If dATP is also included at concentrations below 1 μM, at which it also binds much better to the S site, then the two effectors act jointly to activate the enzyme until the dATP concentration reaches 1 μM. At higher concentrations dATP will start to occupy the A site ($K_{\text{A Site}}^{\text{dATP}}$ = 0.5 μM), resulting in inhibition.

9.3.4 ATP Activates by Binding at the S Site

As shown in Fig. 9.12 when dTTP is absent, ATP activates enzyme activity. Although ATP is normally described as binding at the A site, since the lowest concentration of ATP in this figure is at 200 μM, this would produce significant binding to the S site. Since all the deoxynucleotides that bind at the S site produce activation, it is reasonable to assume that ATP produces its activation also at the S site. Since deoxynucleotides binding at the A site produce inhibition, while ATP does not produce inhibition, a plausible interpretation is that ATP does not stabilize the inhibitory conformation when it binds at the A site.

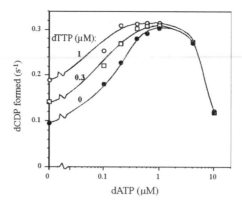

Fig. 9.11. Regulation of dCDP synthesis by dATP at fixed concentrations of dTTP, and the type Ia α_2 dimer at 0.3 nM. dATP up to 1 μM is activating and inhibits when > 1 μM. When combined with dATP, dTTP is an activator. Data from Brown and Reichard[227]

When dTTP is present at a high concentration (50 μM), it has two effects. It inhibits the activity at all concentrations of ATP, and it also changes the enzyme's response to ATP so that positive cooperativity is now observed (Fig. 9.12). This result again demonstrates that several different conformations may be stabilized, as a function of which effector is present at the A site.

9.3.5 Binding at the A Site Determines the Overall Conformation and Activity

Figure 9.13 shows an experiment comparable to that in Fig. 9.10, with dATP at various fixed concentrations. dTTP is an activator only in the absence of dATP. At all concentrations of dATP, dTTP has no effect at concentrations below 10 μM, at which it could bind at the S site. At concentrations of dTTP greater than 100 μM it clearly acts as an inhibitor. Although dATP may also be an activator when it binds at the S site, in Fig. 9.13 it is present at ≥1 μM, and therefore begins to occupy the A site.

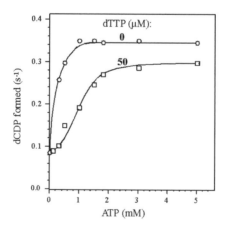

Fig. 9.12. Regulation of dCDP synthesis by ATP at fixed concentrations of dTTP, and the type Ia α_2 dimer at 0.3 nM. ATP at all concentrations activates the enzyme. Data from Brown and Reichard[227]

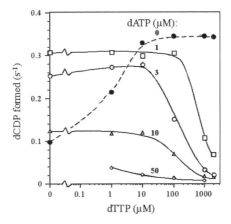

Fig. 9.13. Regulation of dCDP synthesis by dTTP at fixed concentrations of dATP, and the type Ia α₂ dimer at 0.3 nM. In the absence of dATP, dTTP activates. At dATP >1 μM this nucleotide inhibits, and in the presence of dATP, dTTP generally has no effect up to 10 μM, and inhibits at higher concentrations. Data from Brown and Reichard[227]

We see a cumulative effect with two ligands binding at the A site. When dATP concentrations are increased, enzyme activity declines, and combined with increased dTTP concentrations the enzyme is then inhibited to a greater extent. It is not significant that the S site must be saturated with a combination of dATP and dTTP. Only inhibition is evident, demonstrating that the A site controls the final conformation of the enzyme.

There is additional evidence that the conformational effects produced at the A site are not exactly the same as those produced at the S site. The enzyme normally exists as a dimer. When the free enzyme is measured by analytical centrifugation, it has a sedimentation coefficient of 7.8S (Fig. 9.14), and the presence of dGTP or dTTP changes the sedimentation value to 8.2S.[227] Therefore, binding at the S site produces the normally active dimer. In the presence of high concentrations of dATP the sedimentation increases

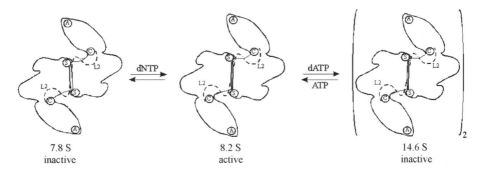

Fig. 9.14. Changes in enzyme activity with change in quaternary structure. Based on the subunit size of the enzyme, the sedimentation values determined are appropriate for the active dimer in the presence of various deoxynucleotides (dNTPs) at 8.2 S, or a conformationally altered dimer at 7.8 S. dATP inhibits and also promotes an inactive tetramer at 14.6 S.[227] The only solved structure is for the active species

to 14.6S, a value consistent with the enzyme having doubled its total size. Addition of ATP prevents this extra aggregation result.

9.4 Regulation of Ribonucleotide Reductase from Calf

In the kinetic studies with the *E. coli* enzyme (above) only a single substrate was tested in detail. To provide a more complete description of the regulatory features, the mammalian enzyme was examined with three of the four possible substrates (Fig. 9.15). This enzyme preparation was not completely pure, and had somewhat lower activity than the bacterial enzyme.[228] Nevertheless, this paper presented the most detailed kinetics attempted with a mammalian preparation of ribonucleotide reductase.

As would be expected, the final end product for a specific substrate does not enhance the activity of the enzyme with the precursor substrate. However, the end product also

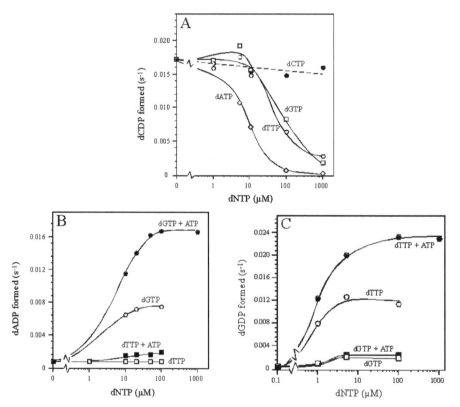

Fig. 9.15. Regulation of ribonucleotide reductase from calf thymus. Activity was measured with the substrates CDP (**A**), ADP (**B**), and GDP (**C**), in the presence of nucleotide effectors. ATP was included at 1 mM for all data points in panel (**A**), and at 0.9 mM where indicated in panels (**B**) and (**C**). Data from Eriksson et al.[228]

provides little inhibition. For the reduction of CDP, dCTP produces a very modest amount of inhibition, even when it is at 1 mM (Fig. 9.15A). And for the reduction of GDP, dGTP actually provides a very slight activation (Fig. 9.15B). We also see that all three deoxynucleotides tested inhibit reduction of CDP (Fig. 9.15A), with dATP being the most effective, as also seen with the enzyme from *E. coli* (Fig. 9.13). The selective activation effects become more evident when ADP and GDP are substrates. With ADP, dTTP has almost no effect, though it greatly activates the enzyme for GDP. dGTP is the better activator with ADP as substrate. With both dGTP and dTTP, the additional presence of ATP increases enzyme activity. This is similar to the ability of ATP to be a general activator with the *E. coli* enzyme.

9.5 Regulation by Alternate, Novel R2 Subunits

In mammals DNA synthesis occurs primarily during the S phase of the cell cycle in dividing cells. It is therefore not essential to maintain high concentrations of ribonucleotide reductase at all times. This provides an additional regulatory mechanism since the gene for this enzyme can be transcriptionally regulated. It has been observed that in mice the beta subunits have a $t_{1/2}$ of only 3 h.[229] Transcription of the gene for the beta subunit is coordinated with the beginning of DNA synthesis, and the concentration of these subunits can be increased by threefold to sevenfold when mouse cells are stimulated to begin cell division.[229]

Humans have two different forms of the beta subunit. The normal subunit, known as hRRM2, is more abundant when cells are going through normal DNA synthesis preparatory to cell division. There is an isozyme of this beta subunit, known as p53R2, that has 80% sequence identity with hRRM1.[230–232] This newly discovered isozyme was named for the fact that the gene coding for it is activated by binding the transcription factor p53. Unlike the gene for the standard beta subunit, the second gene contains a binding site for p53 in intron 1.[232] This makes expression of this gene sensitive to various processes that cause DNA damage: uv radiation, gamma radiation, and chemical agents.

In vitro studies using the hRRM1 subunits (the alpha subunits in previous sections) and either of the beta subunit isozymes showed that the specificity and regulation for reduction of ribonucleotides was the same with either isozyme.[231] One distinctive property is that p53R2 has at least fivefold weaker binding to the alpha subunits, and therefore the holoenzyme formed with p53R2 has about 20–50% lower activity.[230]

The current model for the benefit of having the two isozymes is that the standard beta subunit, hRRM1, is increased in cells that are entering the S phase, and therefore functions chiefly for normal cells that are synthesizing new DNA for cell division. The second beta isozyme, p53R2, appears to be principally used for DNA repair, and thus would be important in all cells. Since in adult mammals genome duplication no longer occurs in most tissues, the appearance of this duplicated gene for the beta subunit led to a change providing a binding site for the transcription factor p53, and thus provided an alternate mechanism for increasing ribonucleotide reductase activity specifically in response to DNA damage.

9.6 A General Model

It is somewhat surprising that in no case did scientists perform the most common enzyme experiment: varying the concentration of the substrate to determine K_m and V_{max}. It would be of interest to have published values for K_m or $K_{0.5}$ for the four normal nucleotide substrates. This reflects the fact that from the earliest studies on this enzyme, the chief focus has been to elucidate the combined action of positive and negative effectors at two separate sites. I include ribonucleotide reductase with the K-type enzymes because the overall activity clearly varies as different effectors are tested, and because we have much evidence that binding of effectors at the S site alters the binding of effectors at the A site. Also, structural studies show how binding of effectors at the S site alters the affinity for substrates at the catalytic site.

In Fig. 9.14 we see that the overall quaternary structure has at least three observed states related to binding at the S site or the A site. Low concentrations of all four deoxynucleotides bind at the S site and promote the active dimer. Some of these deoxynucleotides have been shown to also bind at the A site, where they have much poorer affinity, and where they cause inhibition. dATP binds with the best affinity at the A site, and this leads to the formation of an inactive tetramer. Activity studies suggest that all deoxynucleotides may bind at the A site, causing inhibition. ATP also binds at the A site, but is unable to stabilize the inactive conformation.

Since activation by ATP has been shown with ATP concentrations high enough to also occupy the S site, it is quite possible that activation by all nucleotides occurs at the S site. Because ATP competes for binding with the inhibitory deoxynucleotides, it may be acting as a deinhibitor (activator) at the A site. Positive cooperativity exists for binding to the A sites, when this is the first site to become occupied (i.e., ATP). Negative cooperativity is also evident, since several effectors bind at the S site and more weakly at the A site.

Humans have isozymes for the beta subunits, and increased production of either form is necessary for significant dNDP synthesis. One of the beta isozymes, hRRM2, helps to control nucleotide reduction in mitotic cells, and a second isozyme, p53R2 helps to activate nucleotide reduction when this is needed for DNA repair.

10

HEXOKINASE

Summary

In mammals there are four hexokinase isozymes with different tissue distributions. Because the isozymes have different kinetic features, this has permitted them to become optimally adapted to the needs of the specific tissue in which they are the major isozyme. Isozyme IV (liver) has a subunit M_r of 52 kDa, and has a poor affinity for glucose, but shows positive cooperativity for binding glucose, and this enables it to act as a glucose sensor. This positive cooperativity is completely due to kinetic affects. Isozymes I (brain), II (muscle) and III (erythrocytes) with an M_r of 100 kDa, are the result of gene duplication plus fusion of the smaller form, which gives them the potential to have two binding sites per subunit for each ligand. Evolved changes in such similar sites enable isozymes I and II to both have negative cooperativity for the allosteric inhibitor glucose-6-P, while isozyme II has also developed negative cooperativity for the substrate ATP.

10.1 Introduction

Hexokinases phosphorylate 6-carbon sugars, of which glucose is the most abundant. Many hexokinases also use mannose, fructose, and deoxyglucose as substrates, with somewhat lower affinities. Of the four isozymes found in mammals, hexokinase IV has frequently been named glucokinase, with the implication that it discriminated effectively for this sugar. This name is misleading and should not be used for the mammalian isozyme, since it binds glucose much more weakly than hexokinases I–III (Table 10.1), and shows only a slight preference for glucose.

Many organisms have evolved their carbohydrate metabolism to use glucose as the major sugar, both for storage and for immediate energy production via glycolysis (Fig. 10.1). A significant benefit of using glucose in this role is that among the reducing sugars, it has the lowest activity for reacting with amino residues of proteins to form glycolytic adducts, which become a major problem in diabetes and aging.[233] Since

Table 10.1. Human isozymes of hexokinase

Isozyme:	HK I[a]	HK II[b]	HK III[c]	HK IV[d]
A. Principal Tissue(s)	Brain, muscle Erythrocytes	Muscle Erythrocytes	Erythrocytes	Liver, pancreas
B. Structure				
Subunit M_r (kDa)	100	100	100	52
Oligomer	$\alpha + \alpha_2$	α	α	α
C. Kinetics				
k_{cat} (s^{-1})	100	318	60	62
$K_{0.5}^{Glucose}$ or $K_m^{Glucose}$ (μM)	61	340	38	7,600
$k_{cat} / K_m^{Glucose}$ (M^{-1} s^{-1})	1.7×10^6	0.9×10^6	4.2×10^6	8.2×10^3
K_m^{ATP} (mM)	1.2	1.0	3.05	0.2
$K_i^{Glu-6-P}$ (μM)	15	210	130	No inhibition
P_i blocks inhibition	Yes	No	No	
D. Cooperativity				
Cooperativity for glucose	No	Negative	Positive	Positive
Cooperativity for ATP	No	Negative	No	No
Cooperativity for Glu-6-P	Negative	Negative	No	No

[a] Aleshin et al.[234]
[b] Ardehali et al.[18]
[c] Palma et al.[235]
[d] Gloyn et al.[236]

glucose must be maintained in plasma and other body fluids at ≥ 4 mM, this minimizes the rate of such protein alteration, relative to the other sugars.

Mammals use blood glucose as an immediately available source of energy. Although most of our metabolic energy is derived via mitochondrial oxidation of pyruvate, glycolysis itself may still provide two net ATP per mole of glucose converted to pyruvate. This process may occur in the absence of oxygen, because the organism requires energy at a greater rate than can be provided by oxidation (e.g., skeletal muscle), or because specialized cells have no mitochondria (e.g., erythrocytes). Since only liver and muscle have significant intracellular glycogen stores, other tissues depend on the available steady-state concentration of blood glucose. The four isozymes (Table 10.1) have evolved to meet the differing demands for using glucose by the various tissues.

10.1.1 Functions of Different Hexokinases

It is helpful to know a few values for glucose concentrations to understand glucose metabolism. For mammals normal blood glucose values vary between 4 and 8 mM. For humans, concentrations near 3 mM define *hypoglycemia*, and concentrations above 10 mM define *hyperglycemia*. While all tissues have one or more glucose transporters,

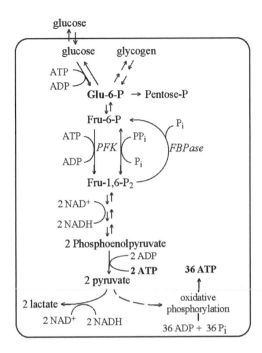

Fig. 10.1. Metabolic pathways that use glucose-6-P

the imported glucose is phosphorylated fast enough that the cellular concentration of free glucose in most cells is normally ≤0.5 mM, which is about 10% of the plasma concentration.[237] Concentrations in muscle are much higher, as shown by the concentration of almost 2 mM for human heart (Table 1.4). Liver has the fastest glucose transporter, and maintains glucose levels similar to those in blood.[238]

Individual isozymes are often named for a tissue from which they are easily obtained, with the unfortunate implication that a specific isozyme is found almost exclusively in the tissue with which it is identified. Tissues identified in Table 10.1 are only meant to help identify an important tissue source, and should not be seen as exclusive. In fact, most mammalian tissues have the three large isozymes (hexokinases I–III) in varying proportions. Liver and pancreas also have hexokinase IV.

Hexokinase I is the main isozyme in human brain and erythrocytes, and since the brain has special features for its energy metabolism (discussed in Chaps. 7 and 8) a unique isozyme may also help. The brain's total dependence on glycolysis plus oxidative phosphorylation is due to its high need for ATP to support ongoing neural function around the clock. A simple quantitative measure of the brain's metabolic status is that the human brain normally accounts for 2% of body weight, but under standard conditions receives at least 20% of the blood supply, to assure adequate oxygen for ATP synthesis.

Consistent with this we see some special features for this isozyme: it shows the best affinity for the binding of glucose (Table 10.1). This guarantees that this isozyme is

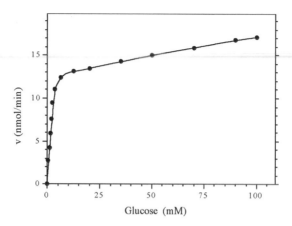

Fig. 10.2. Human hexokinase II has negative cooperativity for glucose. The K_m for glucose is 340 μM. Data from Ardehali et al.[18]

normally near saturation at most intracellular concentrations of glucose in the brain and in erythrocytes, and thereby assures a continued steady rate of ATP production. Hexokinase I also has a high K_m for ATP, and is less sensitive to changes in the concentration of this substrate, so that a partial decline in the concentration of ATP will not significantly diminish formation of glucose-6-P and continued glycolysis.

In humans hexokinase II is an important species in muscle,* but not the major isozyme. Skeletal muscle has the unique need for amplified glycolysis to support stressful work conditions, and therefore a faster enzyme appears to be desirable. This isozyme also has at least three times the activity of isozyme I, which may help when muscles require sudden increased glycolysis during strenuous work. This isozyme shows negative cooperativity for glucose (Fig. 10.2).

The benefit of this negative cooperativity becomes more evident if the intracellular glucose concentration exceeds 4 mM, the concentration in Fig. 10.2 at which the activity curve becomes more constant. However, intracellular glucose concentrations in muscle have been measured at about 2 mM for normal conditions (Table 1.4). Perhaps this isozyme with negative cooperativity is beneficial under conditions of increased glucose import.

Hexokinase III is the least abundant of the isozymes in the body, though it is found in many tissues. It is important in red blood cells, though in human erythrocytes hexokinase I is again the major isozyme,[237] suggesting that these two isozymes are interchangeable. Hexokinase III has about the same affinity for glucose as isozyme I (Table 10.1). The enzyme has modest positive cooperativity for glucose, and is the only isozyme shown to have substrate inhibition at higher glucose concentrations (Fig. 10.3).

---

*
Hexokinase I is frequently defined as the brain isozyme based on its distribution in rats and other organisms. In human muscle hexokinase I is also the major species, at 70% of hexokinase activity, with hexokinase II representing 30%.[246]

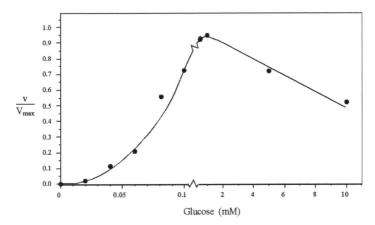

Fig. 10.3. Human hexokinase III has substrate inhibition by glucose. Data from Palma et al.[235]

This has been demonstrated for hexokinase III from pig[239] and from humans.[235] For an erythrocyte enzyme, positive cooperativity for glucose appears beneficial. These cells have little metabolic activity other than the glycolytic pathway, and therefore an appropriate sensitivity to available glucose is assured. Note that the affinity is somewhat high ($K_{0.5} = 38\,\mu M$) for a measured glucose concentration in red blood cells of 0.5 mM.[237]

10.1.2 Continuing Need for a Hexokinase IV

Hexokinase IV has the smaller size consistent with the enzymes in microorganisms, and is therefore closer to a putative common ancestor. While having three 100 kDa hexokinase isozymes that are evolutionarily more recent,[240] mammals continue to use hexokinase IV in the liver and in the pancreas, where it functions as a glucose sensor. In the pancreas the enzyme is involved with activating expression of the insulin gene, and the liver regulates the overall concentration of glucose in the blood. To accomplish this hexokinase IV must be sensitive to changes in the concentration of blood glucose that are normally near 5 mM. This is possible since hexokinase IV has a $K_{0.5}$ of about 7–8 mM.

Let us compare the mammalian hexokinase IV with the yeast hexokinase, since the latter is also a 54 kDa protein, and they probably share a closer ancestor with hexokinase IV. Yeast hexokinase has a specificity constant of $6 \times 10^5\,M^{-1}s^{-1}$, with $k_{cat} = 340\ s^{-1}$ and $K_m = 0.57$ mM.[241] This is near the average value of $4 \times 10^6\,M^{-1}s^{-1}$ for different kinases (Fig. 2.4), and means that yeast hexokinase is fairly normal for kinases. By comparison, the mammalian hexokinase IV has a specificity constant of $8 \times 10^3\,M^{-1}s^{-1}$ (Table 10.1), a value almost 100-fold lower than the yeast hexokinase, and about 500-fold lower than an average kinase. Given that hexokinase IV is an important regulatory enzyme, there must be a compensatory benefit for this otherwise unusually low specificity.

An important constraint is that the liver does not itself depend on glycolysis for its own energy needs, which can be largely met by oxidizing fats. Therefore blood glucose is more important for neural tissues, muscles, and other tissues. The liver enzyme may then function with a much lower activity, while also having an affinity similar to the normal blood glucose concentration. The liver may steadily, and at a low rate, take up glucose to form glucose-6-P when glucose is available, and internally the glucose-6-P will then be mostly converted to glycogen or used in the pentose phosphate pathway (Fig. 10.1). We therefore see an interesting exception to the normal values for specificity that generally balance speed vs. discrimination. The liver hexokinase IV no longer has either.

10.2 Functions of the Duplicated Domain

Hexokinase IV is the smallest isozyme, with a subunit M_r of about 50 kDa.[†] Since the other three isozymes have a subunit size twice as large, the hypothesis of their arising by gene duplication plus fusion of an ancestral glucokinase gene was advanced many years ago, and has been supported by the sequence comparison of the four isozymes.[18, 234] What is the benefit of this gene duplication? Clearly, the initial version of this duplicated protein would simply have had twice as many binding sites for the normal ligands. We have already seen examples with phosphofructokinase (Chap. 8.2.4) where, after some appropriate mutations, this evolutionary process led to an enzyme with altered regulatory functions.

A series of mutations in the rat hexokinase I showed that the native protein has two sites for glucose-6-P, one in the C-terminal half (the catalytic site), and the second in the N-terminal half.[243] The rat hexokinase I was truncated, and studies with the N-terminal half of the enzyme showed no catalytic activity, while the C-terminal half had almost the same activity as the intact enzyme[244] (Table 10.2). However, the N-terminal half has been clearly shown by crystal structures to bind glucose-6-P.[245]

Because for hexokinase I only the C-terminal half had enzyme activity, a widely stated hypothesis emerged proposing that only the C-terminal half had retained an active catalytic center, while the N-terminal half only binds glucose-6-P, and had become a regulatory site with special sensitivity for the product. This new site for inhibition was seen as necessary, since hexokinase IV shows no inhibition by physiological concentrations of glucose-6-P. This also offered an interesting explanation for the appearance of the larger hexokinases, since they now had a new regulatory site.

10.2.1 Product Inhibition at the Catalytic Site

Multiple studies in recent years support a more standard interpretation: product inhibition occurs at the active catalytic site. Several laboratories prepared truncated halves for hexokinase I, II, and III (Table 10.2). Each protein half could then be independently examined for activity and binding. The results with hexokinase I were just described, and show both catalytic activity and normal inhibition with the truncated C-terminal half.

For hexokinase II both halves of the protein have retained catalytic activity.[18] More importantly, both halves show inhibition by the product. A very interesting feature for

---

[†]For hexokinase IV the observed subunit sizes vary from 52 to 59 kDa, due to alternate splicing of the RNA transcript both in liver and pancreatic cells, and this results in proteins with no observed change in function.[242] The M_r value of 50 kDa is widely used.

Table 10.2. Kinetics of the separate halves of human hexokinases I–III

Enzyme	k_{cat} (s^{-1})	$K_m^{Glucose}$ (µM)	K_m^{ATP} (mM)	$K_i^{Glucose-6-P}$ (µM)
Human HK I native	64	53	0.68	37
HK I – C-terminal	46	52	0.50	25
Rat HK I native	100		0.45	14
HK I – N-terminal	0	–	–	14
HK I – C-terminal	93	60	0.9	10
Human HK II native	318	34	1.0	45
HK II – N-terminal	125	46	0.8	59
HK II – C-terminal	76	51	3.8	1,730
Human HK III native	58	34	1.5	23
HK III – C-terminal	60	38	3.0	13

HK I data from Liu et al.[20] and White and Wilson[244]; HK II data from Ardehali et al.[18, 247] HK III data from Palma et al.[235]

this isozyme is that the K_i value for the product is poorer at the C-terminal half (Table 10.2), and that the better K_i at the N-terminal half is very similar to the K_i for the native enzyme. This clearly suggests that binding of glucose-6-P at the N-terminal site is also communicated to the C-terminal catalytic site, where it leads to inhibition by a conformational effect.

With human hexokinase III we also see that the activity of the C-terminal half of the protein is equal to the activity of the native enzyme (Table 10.2). As with hexokinase I, for hexokinase III the K_m and K_i values for glucose and glucose-6-P by the C-terminal alone are about the same as for the intact native enzyme. The binding of ATP by the C-terminal half is weaker. There is again support for interaction between the two halves within a single monomeric protein, since the native hexokinase III has substrate inhibition (Fig. 10.3), while the kinetics with the C-terminal half show no such inhibition over the same range of glucose concentrations.[235]

The hypothesis that the N-terminal half acts as a novel regulatory site had gained support from the observation that hexokinase IV was not inhibited by glucose-6-P (Table 10.1). But, since the product would bind to the glucose site and partly to the ATP site, then the poor binding of glucose to hexokinase IV would naturally result in the product binding very poorly. When K_i values for glucose-6-P were normalized to affinity values for glucose with all known hexokinases from different species[240], it was evident that there was no distinction between the smaller 50 kDa hexokinases and the larger 100 kDa hexokinases. Therefore, the hypothesis that the N-terminal half had evolved as the principal site to mediate inhibition by glucose-6-P is no longer supported by current data.

Information about the three hexokinases may be briefly summarized. The initial duplication created a type II hexokinase with double the normal activity by virtue of two catalytic centers per subunit. But one catalytic center developed much poorer binding for ATP, leading to negative cooperativity for this substrate and an extended concentration range for using this substrate. From this initial duplicated hexokinase two more isozymes evolved (I and III). In these isozymes the N-terminal halves appear not to bind ATP, which accounts for their loss of catalytic activity. The truncated halves of both isozymes can all bind glucose and glucose-6-P. Hexokinase I has the greatest affinity for this product, which helps to regulate the brain isozyme under steady-state conditions. The type III isozyme shows positive cooperativity for glucose, but the benefit for that feature is not yet clear, since this isozyme would be almost fully active at the normal intracellular glucose concentrations. The N-terminal portions have acquired some novel regulatory functions, to be described below.

10.2.2 The N-Terminal Half Has Membrane Binding Sites

In hexokinases I and III the N-terminal half of the protein no longer has a functioning catalytic center. A novel additional function has emerged: the appearance of special binding sites in the N-terminal portion that favor binding of hexokinase to cellular membranes.[246] Both hexokinase I and hexokinase II bind to mitochondria, via a special sequence found in the N-terminal portion.

In carefully designed studies it was possible to follow the rate of glucose phosphorylation by HK I as a function of ATP. Hexokinase activity was greater with ATP generated by oxidative phosphorylation than by equivalent concentrations of ATP that were generated in the reaction vessel by creatine kinase or adenylate kinase, when oxidative phosphorylation was blocked.[247] The position of hexokinase on the outer mitochondrial membrane appeared to give it better access to ATP being formed inside the mitochondrion than to ATP being synthesized in the bulk solvent around the enzyme.

10.2.3 Negative Cooperativity Due to Separate Binding Sites Within a Monomer

Kinetic studies with the truncated halves of isozymes I and II help to explain negative cooperativity in a monomeric enzyme. As an example, with hexokinase II both of the truncated constructs have almost normal activity, but with different affinities for ATP[18] (Table 10.2). This permits the enzyme to have negative cooperativity for ATP, and the two halves on a single enzyme molecule no longer follow the 2-log rule for this substrate, since the binding sensitivity for ATP now extends from 0.08 to 38 mM. This isozyme also shows strong negative cooperativity for the product, glucose-6-P, since the two K_i values vary by almost 30-fold.

10.2.4 Negative Cooperativity Due to a Mixture of Isozymes

It is absolutely standard for scientists to characterize enzymes that are as pure as possible. This clearly avoids any possibility of other enzymes or proteins contributing to the formation of the measured reaction product, or of directly or indirectly interfering with the enzyme itself, or with the detection of product. Kinetic measurements made with such rigorous standards should then provide credible results about the enzyme being examined. However, this approach may miss some of the features that are inherent in having a mixture of isozymes.

Because most cells/tissues have two or more of the hexokinase isozymes, a degree of negative cooperativity must always exist in the kinetics for the hexokinase reaction. Table 10.1 shows that isozymes I and II display negative cooperativity for glucose, or ATP, or the product glucose-6-P. Although the isozymes are all monomeric, such negative cooperativity is possible since they each have a duplicated half at which these metabolites bind, but with a different affinity. In the context of an active cell, the presence of two or more hexokinase isozymes will also provide negative cooperativity because the separate isozymes again display different affinities for substrates and products. This cumulative kinetic pattern is not normally studied, but is clearly inherent in any cellular mixture with multiple isozymes for the same reaction.

10.3 Positive Cooperativity in a Monomeric Enzyme

Hexokinase IV has always received special attention because it displayed positive cooperativity while remaining a monomer. Such positive cooperativity was evident with the principal substrate glucose, and also with two alternate substrates, mannose and deoxyglucose (Fig. 10.4). With fructose as substrate, however, cooperativity was not consistently observed. We see that glucose is the best substrate for hexokinase IV, since both mannose and 2-deoxyglucose (Fig. 10.4) plus fructose (not shown) had much poorer affinities. However, as already defined in Table 10.1, the affinity of hexokinase IV for glucose is almost 100-fold weaker than the affinities of hexokinases I–III. Even though hexokinase IV binds glucose fairly poorly, compared to the other hexokinases (Table 10.1), it was due to this modest discrimination (Fig. 10.4) observed in studies for the enzyme from many organisms that this isozyme is often called glucokinase.

When the rat liver enzyme was finally purified, it was then systematically tested by sedimentation and gel filtration, with protein concentrations of at least 0.25 mg/ml, to assure that the enzyme was sufficiently concentrated to enable oligomer formation. These different studies consistently produced molecular weights between 47,000 and 52,000.[248] Many enzymes may be monomeric in buffer alone, but then form oligomers in the presence of a ligand (normally a substrate) that can stabilize this altered conformation.[25] Therefore Holroyde et al. performed additional tests in which either glucose at 50 mM or ATP at 5 mM was included, but the enzyme remained monomeric.[248] In an additional set of experiments both substrates were included, up to saturating concentrations. The enzyme remained monomeric.[249]

Though such results were clearly seen as unusual, a few other enzymes had also been characterized as monomeric, while kinetic studies demonstrated their cooperativity (Table 10.3). These enzymes have not been characterized as well, but it is assumed that the model for cooperativity for hexokinase IV may apply to other monomeric enzymes.

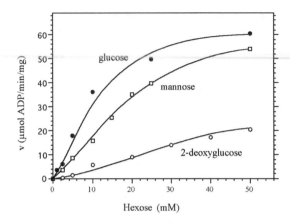

Fig. 10.4. Human hexokinase IV with three hexose substrates. Data from Moukil and van Schaftingen[250]

10.3.1 Models to Explain Cooperativity in a Monomeric Enzyme

The two models described in Chap. 5 for allosteric regulation require the assumption that all subunits in a single oligomer may acquire a more active conformation if only one subunit is binding the ligand that stabilizes the active conformation. Since the empty sites are then in the more favorable binding mode, such ligand binding must produce positive cooperativity. For the monomeric hexokinase IV, a different model is required.

Two models have been proposed that are largely similar (Fig. 10.5), but differ in one important aspect. The *slow transition* model was proposed by Neet and colleagues in 1972,[256] and refined in later papers.[257, 258] The *mnemonic* model was proposed by Buc et al.,[259] and further extended by Cárdenas and colleagues,[260] and by Cornish–Bowden and associates.[261, 262] Let us examine Fig. 10.5 to define the important features for the two models. Although the diagram presents three different structures for the enzyme, if we ignore the low affinity form at the bottom of the figure, what remains represents a normal enzyme with constant affinity: a high affinity form with the same conformation whether empty, or bound with ligand(s), plus a transient species that exists only very briefly as the enzyme goes through the transition state. No crystal structure exists for the true transition state, but a good approximation of such a structure has been obtained with hexokinase IV binding both glucose and an activator.[263] For hexokinase IV itself there is

Table 10.3. Monomeric enzymes that show cooperativity

Enzyme	Organism	Cooperativity	Refs.
Acid phosphatase	Maize	Negative, for p-nitrophenyl-P	251
Hexokinase I	Human	Negative, for glucose-6-P	Table 10.1
Hexokinase II	Human, rat	Negative, for glucose, ATP, and glucose-6-P	Table 10.1
Hexokinase III	Human, rat	Positive, for glucose	Table 10.1
Hexokinase IV	Human, rat	Positive, for glucose	Table 10.1
Ribonuclease	Rat	Positive, for normal substrates	252, 253
Ribonucleotide triphosphate reductase	*L. leischmannii*	Positive, for ribonucleotides	254, 255

also a crystal structure of the enzyme with only glucose, in a different conformation designated "low affinity" and comparable to the *T*-state in allosteric systems.[263] A third conformation with glucose is evident in a crystal structure for hexokinase I.[264] Because hexokinase IV and hexokinase I have sufficient structural similarity, it is expected that hexokinase IV also has this "high affinity" form during some phases of its catalytic cycle.

10.3.2 The Slow Transition Model

The enzyme population is always a mixture of at least two forms: free enzyme in the low affinity conformation and free enzyme in the high affinity conformation, shown in the box at left (Fig. 10.5). If glucose is present then two additional forms of the glucose binding enzyme are in equilibrium (box at right, Fig. 10.5). The key feature of both models is that the enzyme normally exists in a low affinity conformation (the *T* state in the MWC model) that binds glucose very poorly, and is sufficiently stable that the rate for conformational change to the high affinity form must be very slow, relative to the speed of the catalytic cycle. It is equally important that the high affinity conformation does not relax to the low

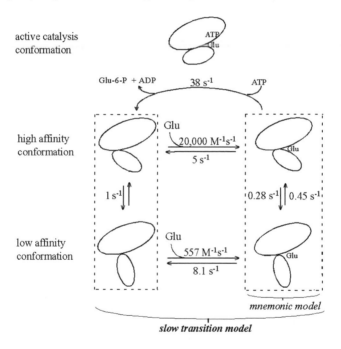

Fig. 10.5. Models for the slow interconversion of hexokinase IV. Data for the human enzyme are from Heredia et al.[265] Conformational equilibria between free enzyme forms, or between glucose-binding forms are shown in *dashed* boxes. The *slow transition* model permits the separate sets of conformational change represented in both boxes. The *mnemonic* model assumes only the conformational changes in the box at right

affinity conformation very rapidly. Once an enzyme has completed a catalytic cycle, the rate for rebinding glucose is much faster $(144\,s^{-1})^{\ddagger}$ than the rate for relaxing to the low affinity conformation $(1\,s^{-1})$.

This slow transition model was initially proposed by Neet and colleagues.[256] Since the enzyme normally has a k_{cat} between 30 and $70\,s^{-1}$, then the rate for relaxation of the high affinity form to the more stable low affinity form should be no faster than $1\,s^{-1}$ so that the presence of substrates may continue to stabilize a conformation that is different from the normal free monomer. Kinetic studies by Heredia et al.[265] demonstrated with fluorescence spectroscopy that tryptophan fluorescence changes as the enzyme binds glucose. They used this approach to follow the conformational change for the glucose-bound enzyme, and determined the rates shown in Fig. 10.5.

In separate binding studies with glucose, the rates of association of glucose to the free enzyme, and the rate of dissociation from the complex were followed by the change in tryptophan fluorescence that results with binding.[266] These studies were done without ATP so that only the binding of glucose was measured. Using (10.1) to model the binding rates, it was calculated that before the addition of glucose only 5% of the enzyme was in

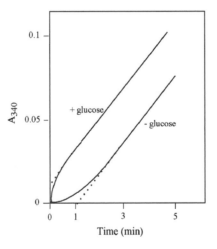

Time (min)

Fig. 10.6. Kinetic burst or lag of hexokinase IV, as a function of the initial enzyme state. For the *upper curve*, showing an initial burst in activity, the enzyme had been stored in 50 mM glucose, and aliquots of this solution were used to start the assay. For the *lower curve*, showing a lag in activity, the same enzyme was extensively dialyzed, before being used to start the assay. For both experiments the reactions had the same concentrations of glucose and ATP. Data from Neet et al.[258]

---

‡The second-order rate constant for glucose binding by the high affinity form was found to be 20,000 M^{-1} s^{-1} (Fig. 10.5). Assuming this to be similar to the enzyme's specificity, k_{cat}/K_m, and multiplying by the K_m value in Table 10.1, the on rate is 144 s^{-1} at saturating glucose (>50 mM). At cellular concentrations of glucose (0.5–0.2 mM) the on rate would be much lower (9–4 s^{-1}). The dramatic positive cooperativity seen in vitro may be partly the result of the high experimental concentrations of glucose being used.

the high affinity conformation, with 95% in the low affinity conformation. This represents a fairly normal distribution between T and R states for an allosteric enzyme.

$$E + \text{Glu} \rightleftarrows {}^*E + \text{Glu} \rightleftarrows {}^*E \cdot \text{Glu} \qquad (10.1)$$

Equally important are the results from this binding study that measured the direct affinity for glucose by the low affinity form and the high affinity form. The K_d for E (low affinity form) was 73 mM, and the K_d for $*E$ (high affinity form) was 60 μM.[266] The actual $K_{0.5}$ for this same enzyme is 7.6 mM (Table 10.1) and this permits the enzyme to then be appropriately responsive to changing values in the actual concentrations of glucose in the blood.

10.3.3 Positive Cooperativity is a Kinetic Effect

10.3.3.1 A Burst or a Lag Defines the Time for Conformational Change

To explore how kinetics may produce cooperativity, studies with rat hexokinase IV demonstrated two different types of progress curves as a function of the enzyme having already bound glucose before being mixed with ATP to initiate the catalytic reaction (Fig. 10.6).[258] The upper curve shows an initial burst, before the curve reaches the evident steady-state position with a constant rate. This burst lasts about 1 min, and is due to the initial high concentration of the glucose-enzyme which has the highest rate for the catalytic step $(38\,\text{s}^{-1})$[§] as it binds ATP as the second substrate and completes catalysis. Since the experimental concentration of glucose in the reaction is only 1.0 mM, all molecules do not rebind glucose after the initial catalytic cycle, and therefore some of these relax to the low affinity conformation. The steady state then reflects the smaller population of enzyme that is able to rebind glucose at the available concentration of 1.0 mM. To perceive the rationale for the two concentrations of glucose used in this study, note that hexokinase IV is nearly saturated with glucose at 50 mM, but due to its high $K_{0.5}$ it has very low activity with glucose at 1.0 mM (Fig. 10.4).

The lower curve (Fig. 10.6) shows the result when free enzyme is used to initiate the reaction. Initially there is very little activity, and the time until the progress curve reaches the steady-state position is the lag. For the conditions of this experiment the lag was just over 1 min. This experiment clearly shows that the free enzyme is not in the high affinity conformation, and due to the slow binding of glucose by this conformation, some time is required before enough molecules enter the catalytic cycle, at the end of which they remain in the high affinity conformation to produce the steady-state progress curve.

10.3.3.2 Ligand Binding Itself Has No Cooperativity

The other important point about the rates for the conformational change is that they are sufficiently slow that by this kinetic feature at least two subpopulations of the monomeric

---

[§]Liver hexokinase IV preparations from rat were often unstable, and normally had much lower kinetic values than the human enzyme. For consistency, kinetic values for the human hexokinase IV will be used in these comparisons.

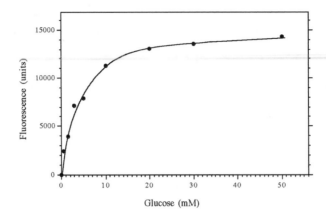

Fig. 10.7. Equilibrium binding of glucose by human hexokinase IV. The K_d is 4.5 mM, and $K_{0.5}$ is 7.2 mM for the same enzyme. Data from Heredia et al.[265]

enzyme are possible, and their ratio may be changed by the presence of substrates. In classic models for allosteric control, ligand binding produces cooperativity by stabilizing a conformation in an oligomer that then has properly formed, but empty binding sites. Such cooperative effects are seen in binding curves for Hb–O_2 (Fig. 6.2).

By following the change in the intrinsic fluorescence of the human hexokinase IV as it binds glucose, it was possible to directly monitor the binding of glucose itself, since no ATP was present to complete the catalytic reaction. These results (Fig. 10.7) show that the direct binding of glucose to the free enzyme follows a normal hyperbolic binding curve, and demonstrates that no cooperativity is observed.[265] The binding experiment was at equilibrium, therefore the kinetic rates shown in Fig. 10.7 do not influence the distribution of the two conformations. This is a very significant experiment, since it demonstrates that the cooperativity seen with hexokinase IV is only accomplished by a protein with a structure that undergoes conformational changes slowly. Since the catalytic rates are sufficiently faster, then a fraction of the monomeric enzyme may always be kinetically trapped in the less stable, but higher affinity conformation. Because the end result of such a kinetic mechanism is positive cooperativity, hexokinase IV has a much more sensitive response to available glucose concentrations, even though the midpoint of this response curve, the $K_{0.5}$, is at a fairly high value.

10.3.3.3 Glucose Activates Phosphorylation of Fructose

One additional set of experiments demonstrates the kinetic nature of the positive cooperativity for hexokinase IV. Fructose is a poor substrate due to its lower affinity. Low concentrations of glucose produce an almost fivefold increase in the enzyme's activity with fructose (Fig. 10.8). As the concentration of glucose approaches the K_m for

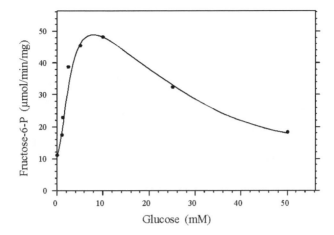

Fig. 10.8. Activation of human hexokinase IV by glucose, when fructose is the measured substrate. Although glucose is also a substrate, the assay measures fructose-6-P. Data from Moukil and van Schaftingen[255]

glucose, it is no longer an activator because the enzyme now preferentially uses glucose as the substrate, which appears as inhibition (Fig. 10.8).

This increase in activity for the phosphorylation of fructose must result by the same kinetic mechanism described above. As glucose binds to the enzyme it may also itself be converted to product. Nevertheless, by having an enzyme molecule go through the catalytic cycle, this enzyme molecule briefly remains in the high affinity form and then may bind either fructose or glucose for the next catalytic cycle. Since glucose has much better affinity than sucrose, it promotes a greater number of enzyme molecules to the active form, compared to a fixed concentration of fructose.

10.3.4 The Slow Transition Model Vs. the Mnemonic Model

The key distinction of the mnemonic model is that it focused on conformational changes of the glucose-bound form of hexokinase IV (box at right, Fig. 10.5).[268] This may have been due to the observation that the T form of hexokinase IV has very poor affinity for glucose and therefore very poor activity. The early work on defining positive cooperativity was largely done with impure preparations of rat liver hexokinase IV, which were somewhat unstable and had modest to poor activity.[256, 249, 267, 268] It is only in the more recent studies with the purified human enzyme that consistent kinetic values have been obtained.[265] These studies demonstrate that the slow transition model gives a better description by emphasizing that the observed cooperativity is provided completely by the kinetic transitions. The main proponents of the mnemonic model, Cornish–Bowden and Cárdenas, eventually accepted the slow transition model as the more general description for the cooperativity of hexokinase IV.[262]

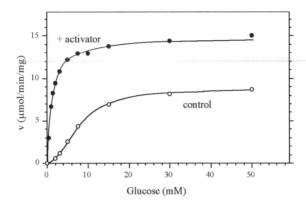

Fig. 10.9. A synthetic activator for human hexokinase IV stabilizes the high affinity conformation. $K_{0.5}$: control, 8.0 mM; plus activator at 30 µM, 0.6 mM. Data from Kamata et al.[263]

10.4 Regulators of Glucokinase

Important regulatory enzymes are normally subject to activation as well as inhibition. Efforts have therefore continued to discover and characterize such effectors for hexokinase IV.

10.4.1 Small Molecule or Metabolite Activators

No normal cellular metabolite has so far been identified as an activator of hexokinase IV. When researchers screened libraries of available chemical compounds to detect activators for hexokinase IV, two different compounds were detected that bind the enzyme with an affinity in the low micromolar range and also produce significant activation.[263, 265] The activator binding site is well defined in the crystal structure, and occurs in a region of the enzyme that has received interest because numerous human mutations located there have produced enzymes with higher activity[130]. Such mutations were easily discovered because the carriers have chronic hypoglycemia.

The effect of one activator designated compound A, on human hexokinase IV is shown in Fig. 10.9. While the effect of the activator is largely K-type, since the affinity of the enzyme for glucose increases tenfold, there is also an increase of 50% in V_{max}. This activator was used in preparing crystal structures of the human enzyme, so that the binding site is well defined. The enzyme form binding the activator is the high affinity form (Fig. 10.5). In the structure of the low affinity form the binding site for the activator is absent due to the conformational change. This demonstrates that the activator functions by stabilizing the high affinity conformation of the enzyme. Note that this is not simply a kinetic effect, since there is an independent binding site for this activator, so that it may stabilize the active conformation as described in the MWC and KNF models.

Cryptic and undiscovered allosteric sites were discussed in Chap. 4.2.2. The site where these different activators bind on hexokinase IV may then represent a true allosteric site, for a physiological activator yet to be discovered, or it may be a cryptic site for which two appropriate synthetic analogs have good binding affinity.

10.4.2 Glucokinase Regulatory Protein

Hexokinase IV is also inhibited by a regulatory protein, which acts competitively with glucose for binding to hexokinase IV.[269] This 68 kDa protein is itself allosteric, and maintains the active conformation, as an inhibitor of hexokinase IV, when it binds fructose-6-P (Fig. 10.10). When fructose-6-P is replaced by fructose-1-P, no inhibition of hexokinase IV is observed.[238, 270] The metabolic rationale for these control features comes from the ability of the liver to use free sorbitol and fructose to produce fructose-1-P. If fructose-6-P is abundant, then synthesis of glucose-6-P is not as necessary, and therefore it is practical to shut down hexokinase IV. Therefore fructose-6-P represents active synthesis of glucose-6-P, and acts to inhibit hexokinase IV via the regulatory protein. Fructose-1-P represents an alternate dietary source for glycolytic intermediates and gluconeogenesis, and therefore acts to de-inhibit hexokinase IV by stabilizing the inhibitory protein in a nonbinding conformation.

10.4.3 Inhibition of Hexokinase IV by Lipids

Fats are a good source of calories, and scientists explored the possible inhibition of hexokinase IV by fatty acids. If energy from fatty acids is available, then for the liver there there is less need for glycolysis. Using rat hexokinase IV it was shown that various long chain acyl-CoAs show very effective inhibition. C-18 acyl-CoAs had K_i values below 1 μM, while C-16 and C-12 acyl-CoAs had somewhat higher K_i values from 2 to 14 μM.[271] The acyl-CoAs acted as competitive inhibitors vs. both ATP and glucose. This is consistent with these molecules binding to the cleft where the normal substrates bind. The free fatty acids themselves showed no inhibition, when tested up to 1 mM.

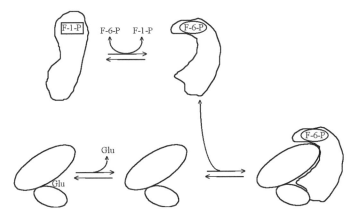

Fig. 10.10. A hexokinase IV regulatory protein. The regulatory protein binds fructose-6-P and fructose-1-P, possibly at the same site. The two ligands stabilize different conformations. The active conformation binds to the free form of hexokinase IV as an inhibitor. Glucose competes with the regulatory protein. No structure exists

Although these results have been criticized,[238] they may be valid since the authors worked with fairly pure enzyme, and performed extensive studies with both the free fatty acids, which had no effect, and the acyl-CoAs which showed fairly good inhibition. The K_i values determined are good enough that cellular concentrations of these acyl-CoAs could have some regulatory effects.

10.4.4 Need for Multiple Hexokinases

Do we actually need four hexokinase isozymes? At this point no consensus physiological rationale or model has emerged, but we can consider the properties for the different hexokinases (Table 10.1) and see some possible benefits while comparing all four enzymes for their activity with glucose (Fig. 10.11). Hexokinase I has the highest affinity for glucose. Hexokinase II has negative cooperativity, reflecting that it contains two separate catalytic domains with different affinities for glucose. Hexokinases III and IV show positive cooperativity for glucose, a kinetic effect described in Chap. 10.3.2. Hexokinase III is also inhibited by high concentrations of glucose, while hexokinase IV has the poorest affinity for glucose.

The benefit for isozymes is evident in a physiological context (Fig. 10.12). Hexokinase I is the main isozyme in brain, and since the brain has special features for its energy metabolism (see Chaps. 7 and 8) a unique isozyme may also help. The brain's total dependence on glycolysis plus oxidative phosphorylation is due to its high need for

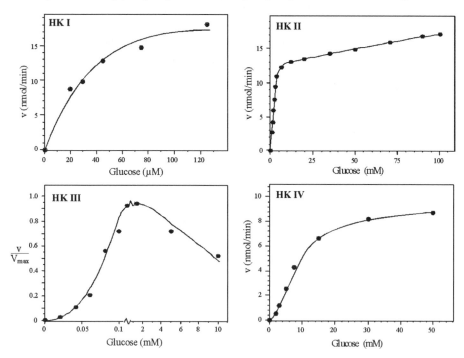

Fig. 10.11. The four human hexokinases. Note the difference in glucose concentrations

ATP to support ongoing neural function around the clock. Consistent with this we see the principal feature for the brain isozyme: it has a very high affinity for glucose, so that the enzyme will be near V_{max} for most intracellular glucose concentrations in brain or red blood cells. These tissues are highly or completely dependent on glycolysis for energy, and the kinetics of HK I assure that the formation of glucose-6-P and glycolysis will not be significantly diminished. For the same reason, this isozyme has a steady response to glucose-6-P, and an increase in its formation will not significantly alter the enzyme's rate.

Skeletal muscle has the unique need for amplified glycolysis to support stressful work conditions, and therefore a faster enzyme appears to be desirable. In humans hexokinase I and II are the main isozymes in muscle. Since hexokinase II has an almost tenfold higher K_m for glucose, the presence of these two isozymes gives muscle a wider range of responsiveness to changes in the muscle glucose concentration (Fig. 10.11). Hexokinase II with the poorer affinity will be sensitive to normal muscle glucose concentration, and isozyme I will be fairly active at lower glucose concentrations should hypoglycemia occur. HK II also has more than threefold the activity of isozymes I and III, which may help when muscles require sudden increased glycolysis during strenuous work. It is perhaps for this reason that hexokinase II also shows negative cooperativity for glucose (Fig. 10.2).

10.5 Evolution of Hexokinases

Hexokinase IV is the smallest mammalian isozyme, with a subunit M_r of about 50 kDa. An interesting problem remains, since even the 50 kDa form of the enzyme is remarkably large for an enzyme with a normal phosphotransfer activity. Similar to the majority of enzymes (Fig. 1.5), many kinases have subunit sizes below 30 kDa, which is appropriate for the size for a single domain enzyme. Consistent with this size expectation, a number of hexokinases from microorganisms have a subunit size at 24–25 kDa (reviewed by Cárdenas et al.[240]). Although the 50 kDa enzymes have no obvious internal sequence duplication,[240] it has been suggested that they also evolved by duplication plus fusion of a smaller 24–25 kDa precursor. This would be consistent with a second such duplication plus fusion to produce the many 100 kDa isozymes, but this hypothesis lacks clear support.

10.5.1 Expansion of a Basic Precursor

However, different microorganisms also contain hexokinases at sizes of 30–42 kDa,[272] so that we almost have a continuum of intermediate values from 24 to 50 kDa. Such a size range could occur if in some species an extra function were provided by incorporating one or more modules with an average size of 5 kDa or even an extra domain of about 20 kDa. Where such size increments provided an extra ligand binding feature that added an important control function for the enzyme's activity, we may then consider such size expansion as being evolutionarily beneficial.

Unfortunately, none of these enzymes with subunit sizes that are intermediate between 24 and 50 kDa have been well characterized, so that no specific benefits are

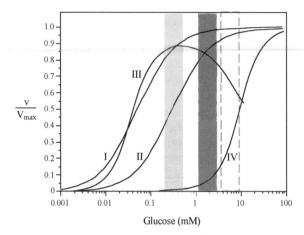

Fig. 10.12. The four human hexokinase isozymes. Intracellular glucose concentrations: *light shading* – low (brain and other tissues), *dark shading* – moderate (muscle), *dashed box* – similar to blood (liver). Compare these *curves* to the same curves shown individually on a linear scale in Figs. 10.2–10.4, 10.9, and 10.11

currently characterized as evolutionary advances. However, for the 50 kDa mammalian enzymes, two novel functions have recently been described, and lend support to this having occurred in an earlier precursor. The discovery of a cryptic activator site was described above, and it is likely that an appropriate physiological molecule will soon be identified as the ligand that binds here. If such a physiological control agent is identified, it will support the hypothesis that this should have been an early evolutionary advance, suggesting that an ancestral glucokinase of 25–30 kDa acquired an extra binding site by incorporating the protein fold for such a site.

We have already examined the bacterial phosphofructokinase (Chap. 8.2.2), and see that it is a sophisticated allosteric regulatory enzyme, although it has a subunit size of 32 kDa. Therefore, we may expect that expanding a hexokinase to a size of about 35 kDa may have produced the earliest form with a regulatory site. It has also been observed that hexokinases I and II are largely associated with the mitochondrial outer membrane.[273] Whether free in solution or associated with the mitochondrial membrane, hexokinase I has the same kinetics.[274] One benefit of this localization would be to assure the enzyme of ATP immediately as it is produced by the mitochondrion.

Since only hexokinase II has catalytic activity in both halves of the protein, it may best represent the ancestral gene duplication of a glucokinase, since that original event should have produced a protein with two functioning catalytic sites per subunit. For hexokinases I and III, mutations leading to loss of activity were compensated by better binding of the enzyme to cellular membranes. However, the ancestral glucokinase may well have had positive cooperativity for glucose, and that feature now remains only with the type III hexokinase, and with hexokinase IV.

SECTION 3

V-TYPE ENZYMES

11

V-TYPE ENZYMES

Summary

V-type enzymes are named for their often remarkable changes in V_{max} when they are allosterically activated. However, these enzymes also generally have a comparable change in their affinity for the substrate. When such enzymes are in a normal metabolic pathway, changes in the maximum activity can be up to 30-fold. To be included in this category, enzymes must have at least a doubling in V_{max}, so that this allosteric effect can be physiologically or experimentally meaningful. When these enzyme activities are part of a regulatory switch, the change in activity may exceed 100,000-fold.

11.1 Overview of *V*-Type Enzymes

The division of regulatory enzymes into subcategories of *K*-type and *V*-type is clearly somewhat arbitrary. There are no thermodynamic constraints that enzymes can have a change only in their affinity or only in their maximum activity. The actual data show that most *V*-type enzymes normally change both (Table 11.1), and one could define them as *V*+*K*-type enzymes, though such a description is almost never used. Also, it may be argued that there can only be one true V_{max}, the highest activity measured at saturating substrate and saturating activator concentrations. The name "*V*-type" serves mainly to emphasize the feature that researchers consider to be more important as the regulatory mechanism or for the enzyme's function.

For the many *K*-type enzymes the substrate itself commonly stabilizes the more active conformation in the *T*:*R* equilibrium, as indicated by the sigmoid kinetic curves. Since saturating substrate can itself produce the highest activity, it is only the change in affinity that is emphasized, leading to the definition of such enzymes as being *K*-type. Perhaps a more useful distinction to define *V*-type enzymes is that the substrate does not itself stabilize the more active conformation. Therefore, even at saturating substrate the

Table 11.1. V-type enzymes

Enzyme (species)	Effector	V_{max} Control	V_{max} + Effector	ΔV_{max}[a]	K_m (μM) Control	K_m (μM) + Effector	ΔK_m[a]	Refs.
Small change in V_{max} and K_m								
Aldose reductase (rat)	H_2O_2	1.2 mU	2.3 mU	0.92	83.6	93.4	0.12	275
ATPase (bovine)	Tropomyosin	2.3 s^{-1}	6.2 s^{-1}	2.7	50	50	0	276
Mercaptopyruvate sulfurtransferase (rat)	Thioredoxin	9.3 s^{-1}	21.7 s^{-1}	2.3	500	240	2.1	277
Uracil DNA glycosylase (*E. coli*)	SSB[b]	34 s^{-1}	10.8 s^{-1}	3.3	0.6	0.5	0.16	278
Uracil phosphoribosyltransferase (*S. solfataricus*)	UMP	0.7 s^{-1}	3.8 s^{-1}	5.4	179	89	2.0	135
DnaK chaperone ATPase (*E. coli*)	Peptide	0.0005 s^{-1}	0.0153 s^{-1}	30.7	0.052	14.6	280	279
Guanylate cyclase (bovine)	•NO	0.79 s^{-1}	700 s^{-1}	886				280
Larger change in V_{max} and K_m								
AMP nucleosidase (*A. vinelandii*	ATP	0.05 s^{-1}	20 s^{-1}	400	6.0	0.1	60	281
P21*ras*	GAP	1.2×10^{-4} s^{-1}	19 s^{-1}	158,000	0.0005	9.7	19,400	282
Unknown change in K_m								
3-Phosphoglycerate dehydrogenase (*E. coli*)	Pyruvate	4.5 s^{-1}	113 s^{-1}	25.1				283
Pyruvate carboxylase (chicken)	Acetyl-CoA	2.1 mU	120 mU	57.1				284

[a] The change in multiples of the control value is shown
[b] Single stranded binding protein

V_{max} observed must be lower than the V_{max} in the presence of one or more activators. V-type enzymes that have the more dramatic change in V_{max} normally function as regulatory switches, so that one conformation has an unusually slow activity, which can then be increased a thousand fold or more, but still be a modest k_{cat}. A good example of this is the G protein p21*ras* (also discussed in Chap. 12), which, after an increase in V_{max} of over 150,000-fold, has an activity of only 19 s^{-1}.[282]

The fact that the change in affinity is in a comparable range to the change in activity is evident in Fig. 11.1 where the data from Table 11.1 are plotted. Over a range of five orders of magnitude there is a very good correlation between the change in V_{max} and the concomitant change in K_m. We therefore see that all allosteric enzymes are K-type, and that a subset has the additional property of an altered V_{max}. This duality is not normally appreciated, or at least not normally discussed.

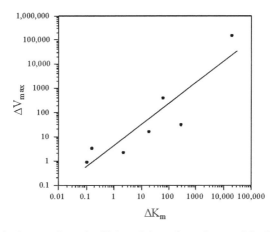

Fig. 11.1. The correlation between change in affinity and change in maximum activity for *V*-type enzymes. Data are for the enzymes in Table 11.1[285]

11.2 Enzymes with a Modest Change in V_{max} and K_m

11.2.1 Mercaptopyruvate Sulfurtransferase

This enzyme is involved in the detoxification of cyanide by transfer of a sulfur group (cyanide + 3-mercaptopyruvate → thiocyanate + pyruvate), and this leads to oxidation of cysteines on the enzyme. The enzyme exists in a monomer:dimer equilibrium, with dimers being stabilized by intersubunit disulfide bonds. This occurs whenever cellular oxidizing conditions are increased.[277] The reduced monomer is more active, while the oxidized dimer has lower activity.

If the enzyme is reduced by thioredoxin, then activity is increased by twofold to threefold. Reduction of the disulfide bond by thioredoxin converts the dimer to the more active monomer. This change is not seen as a normal regulatory mechanism, but simply as a consequence of the enzyme's cysteine being sensitive to oxidation. The change in oligomerization results in a change not just in activity, but also in affinity. While the process for conformational change is different, the change in affinity is again comparable to the change in k_{cat} (Table 11.1, Fig. 11.1).

11.2.2 Uracil Phosphoribosyltransferase

Uracil phosphoribosyltransferases were discussed in Chap. 4.3. The enzyme from *S. solfataricus* has very evident changes in V_{max} when activated by GTP or when inhibited by UMP (Fig. 11.2). GTP is an activator, because it represents purines, and therefore stimulates the production of UMP by this enzyme, since UMP is a precursor of both pyrimidine nucleotides UTP and CTP. As GTP concentrations approach 300 μM activation becomes optimum. This value is still below the average cellular concentration of 450 μM, so that GTP should be an activator under most conditions.

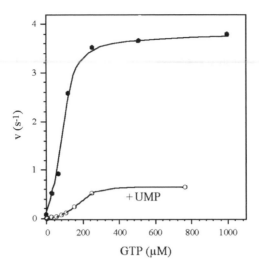

Fig. 11.2. Activation of uracil phosphoribosyltransferase by GTP and inhibition by UMP. Data from Jensen et al.[135]

Cooperativity for GTP may not be obvious in this curve, but becomes very evident when a fixed concentration of UMP is included. The product UMP is a natural inhibitor, and binding of UMP not only lowers overall activity by simple competitive inhibition, but also has an added effect on the GTP binding site since we now see that the UMP-bound form of the enzyme has a poor affinity for GTP. The optimum position of the lower curve (Fig. 11.2) is more representative of cellular conditions, in that both nucleotides are present. In the presence of UMP the binding of GTP is modestly improved as that activator increases in concentration.

At least three different conformations for uracil phosphoribosyltransferase are evident in these data. In the absence of GTP and UMP the enzyme has a very low activity of about $0.4 \, s^{-1}$. Physiological concentrations of GTP produce a more active conformation, and improve this activity by nearly 20-fold. The product UMP at 1 mM (a concentration somewhat higher than cellular) dramatically inhibits the enzyme, consistent with its very good K_i of 0.5 μM. UMP must also change the conformation of the enzyme since it results in much stronger cooperativity for GTP.

The pattern that we see for the affinities of the activator and the inhibitor are again normal, since the enzyme is more inhibited (or less active) at standard concentrations of these two nucleotides, but can become more active in the production of UMP as either UMP itself decreases in concentration, or as GTP increases. We see then that these different conformations not only result in dramatic changes in V_{max}, but in changes in affinity almost as great (Table 11.1).

11.2.3 Phosphoglycerate Dehydrogenase

Phosphoglycerate dehydrogenase is the committed and rate limiting step in the synthesis of serine. The metabolic sequence oxidizes the glycolytic intermediate 3-phosphoglycerate to hydroxypyruvate, which is then aminated, and subsequently dephosphorylated to produce serine. As the end product of this pathway, serine is a feedback inhibitor on the initial enzyme, phosphoglycerate dehydrogenase. In addition, since the first substrate is an intermediate in glycolysis, pyruvate is an activator. An abundance of pyruvate represents a well nourished or high energy state for bacteria in which this enzyme occurs, and therefore favors the use of 3-phosphoglycerate for the synthesis of an amino acid.

If we examine first the activation of the enzyme by pyruvate (Fig. 11.3A) we see that in the absence of pyruvate the enzyme still has a very modest activity of about $4.5 \, \text{s}^{-1}$. As the concentration of pyruvate is steadily increased the observed activity increases by 25-fold. Note that the response of the enzyme to pyruvate shows a classic hyperbolic curve, indicating a constant affinity by the enzyme for this activator. While such an in vitro activation curve looks impressive, the K_A for pyruvate is about 6 mM. Since the intracellular concentration of pyruvate is not normally above 1 mM, pyruvate will normally not have a significant activating effect. However, it is possible that *E. coli* (the source for this enzyme) experience sufficient fluctuations in available carbohydrate precursors that when the colon receives the remnants of a meal the bacteria are prepared to utilize this brief time period for a sudden burst of protein synthesis and cell duplication.

Inhibition by serine (Fig. 11.3B) is much more dramatic because the K_i for serine is so much lower. Inhibition is nearly total when serine is at a concentration of 25 μM. Since amino acid concentrations are usually above this value, then this pathway should have only minimal activity under most conditions. The affinity values for the activator

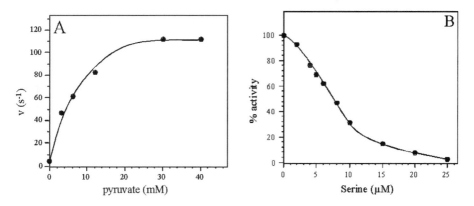

Fig. 11.3. Allosteric effectors of the phosphoglycerate dehydrogenase reaction. (**A**) Pyruvate raises the maximum observed activity by 25-fold. (**B**) Serine decreases this activity to a comparable extent. Data from Sugimoto and Pizer [283]

and the feedback inhibitor are consistent with this pathway functioning only under special conditions: when the cell experiences a surplus of carbohydrates, leading to high pyruvate levels, or when the cell has a deficit of the amino acid serine, leading to loss of the feedback inhibition.

Inhibition is clearly more important as a regulatory mechanism. This is evident in the stronger affinity for the inhibitor, and also in the cooperativity shown toward serine. This cooperativity permits the enzyme to be almost insensitive to the inhibitor when serine concentrations are low, but then respond with more dramatic inhibition as the concentration of serine reaches slightly higher levels.

11.3 Enzymes with Large Change in V_{max} and K_m

11.3.1 AMP Nucleosidase

AMP nucleosidase catalyzes the irreversible cleavage of the glycosidic bond to produce adenine and ribose-5-P. As AMP is a precursor for ATP, the normal need for adequate ATP concentrations requires that this enzyme should have no or little activity.

In the absence of ATP, the activity is only $0.05\ s^{-1}$ (Table 11.1). By having ATP function as the activator for this enzyme, removal of AMP will occur only when ATP concentrations are high enough for the cell, and lead to a new V_{max} of about $20\ s^{-1}$. The surplus AMP can then be cleaved and the ribose-P recycled for the synthesis of other nucleotides.

Since the enzyme shows positive cooperativity for ATP (Fig. 11.4A) this again dampens the activation effect when concentrations of ATP are low. Only as concentrations of free ATP approach 100 μM is optimum activity achieved. These results for an enzyme from the bacterium *Azotobacter vinelandii* are consistent with earlier

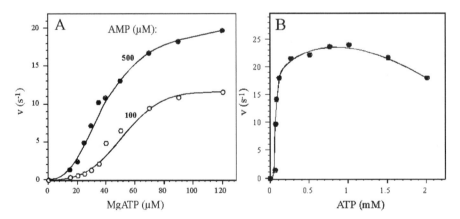

Fig. 11.4. Activation of AMP nucleosidase by ATP. Inhibition occurs by competition at the catalytic site. Data from Schramm[281]

discussions suggesting that the free ATP concentrations in most cells are normally less than 500 μM (Chap. 2.3.2). Since the $K_{0.5}$ for AMP is 110 μM, the curve at 500 μM AMP (Fig. 11.3A) shows the normal increase in activity with this increase in the concentration of the substrate.

Concentrations of ATP above 1 mM lead to inhibition (Fig. 11.4B). This inhibition most likely occurs by ATP binding at the catalytic site, and thereby blocking access by the normal substrate AMP. Activation occurs at low concentrations of ATP, with a $K_{0.5}$ of 48 μM.[281] The much higher concentrations required for inhibition by ATP show that ATP could bind at the catalytic site as an analog of AMP, but with much poorer affinity. Since cellular concentrations of free ATP are not likely to reach 1 mM, this inhibition should not be significant in vivo.

11.3.2 Regulatory Switches

The most dramatic examples of increases in V_{max} are by enzymes such as the G proteins with dramatic increases in their normally slow GTPase activity, and also protein kinases in signaling cascades. For these we observe rate enhancement of many thousand fold. The G protein p21*ras* will be described in Chap. 12, and protein kinases are covered in Chap. 13.

11.3.3 Exceptions

Not all enzymes reported as *V*-type have a true change in V_{max} as discussed here, where we have considered pure enzymes that respond to a known effector binding at a regulatory site. Examples where the total cellular activity of an enzyme increases can be represented as a higher V_{max}, since the total cellular protein does not change much. An example of this is in a report that acetyl-CoA carboxylase activity has a greater V_{max} in cell extracts, after cells have been exposed to insulin or to epidermal growth factor.[286] However, since the increase in the V_{max} was only twofold, this could also have resulted by the known effects of such agents in stimulating specific gene transcription, leading to an increase in the total amount of those enzymes, rather than a change in their actual V_{max}.

12

G PROTEINS

Summary

G proteins such as p21*ras* are designed to hydrolyze the bound GTP very slowly, so as to maintain a special conformation that enables them to bind and activate various cellular receptors and protein kinases. A GTPase activating protein (GAP) can increase this hydrolytic step by up to five orders of magnitude. For this feature these enzymes are called *V*-type enzymes, even though many G proteins also have significant changes in K_d for the bound substrate or allosteric effector. GAP is an inhibitor, since by activating the GTPase rate it limits the time for the protein to be active. Guanine nucleotide exchange factors (GEFs) lower the affinity for the tightly bound GDP and promote replacement by GTP.

12.1 Overview of G Proteins

G proteins are so named because their conformation is regulated by binding either GTP or GDP. In the active conformation, with GTP bound, these proteins act as effectors able to bind many receptors and activate their physiological function (Fig. 12.1). These functions include adenyl cyclase, phosphodiesterases, phospholipases, and ion channels.

12.1.1 p21*ras*

The key feature of such signaling proteins is that the active form, E·GTP, has a limited lifetime as its intrinsic GTPase activity slowly hydrolyzes the GTP to GDP. The p21*ras*·GDP that is formed changes conformation (Fig. 12.1), and is no longer active in binding target enzymes. p21*ras* has a subunit size of 21 kDa and is designated as a *V*-type enzyme because a GTPase activating protein (GAP, Fig. 12.1) acts as an allosteric effector, and increases the rate for GTP hydrolysis by about 150,000-fold (Table 12.1). It must be noted that the change in affinity for GTP produced by GAP (about 19,000-fold) is also very large.

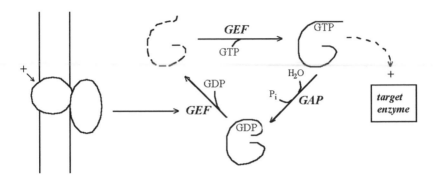

Fig. 12.1. Interconversions of G proteins such as p21*ras*. p21·GTP is the active form, and binds to and activates various cellular target enzymes. Slow hydrolysis of GTP leads to the inactive p21·GDP, and thereby limits the time for activation by p21·GTP. GAP (GTPase activating protein) is an inhibitor since it greatly activates GTPase activity. GEF (guanine nucleotide exchange factor) is an activator since it promotes release of GDP and binding of GTP, and is produced by various signals that can act through membrane receptors. The unliganded p21 is very unstable

A continuing difficulty with the study of G proteins and other switches is that values measured in vitro with purified protein frequently are not consistent with observed rates in vivo. Studied in cells, the human p21*ras*·GTP has a $t_{1/2}$ of about 2–3 min.[287] When this slow rate for the GTPase activity is measured in vitro, a k_{obs} of about $1.2 \times 10^{-4} s^{-1}$ was determined.[282] Since the in vitro rate is comparable to one catalytic event per 139 min, other factors must operate in vivo to account for this difference, and these results led to the search for the exchange factors (GEFs, Fig. 12.1).

12.1.2 Mechanism of p21*ras*

The overall control of the conformational changes comes from the differences in the binding lifetime for GDP vs. GTP. Let us consider the cellular cycle for such proteins. Under normal conditions p21 is bound with GDP, a ligand for which it has a very high affinity of 22 pM.[288] This assures that p21·GDP exists as a very stable species, since the off rate for the GDP is 2.8 h.[288] Since p21·GDP has the inactive conformation, then this protein must be stimulated by some physiological change that leads to expression of the guanine nucleotide exchange factors (GEFs). GEFs bind to p21 and favor the release of the tightly bound GDP, which permits binding of GTP (Fig. 12.1). The newly formed p21·GTP is now the active protein, and can bind with cellular targets. There are many different types of such G proteins, each normally specific for a subset of enzyme targets. p21*ras* is the most extensively characterized and will be our main focus.

The duration for this initiated activation process is limited by the time it takes for the bound GTP to be hydrolyzed, causing the protein to return to the p21·GDP state with no activity. This slow internal hydrolysis of the activating nucleotide thus acts as a clock or timing device to assure that the activation signal only lasts for that length of time that is dictated by the GTPase rate.

Table 12.1. GTPase rates of p21*ras*

Enzyme	k_{cat} (s^{-1})	Δk_{cat}	K_m	ΔK_m	Refs.
p21*ras*	1.2×10^{-4}		0.5 nM		282
+ GAP	19	158,000	9.7 µM	19,400	

The enzymes discussed in Chap. 12.1 and 12.2 are controlled either by the binding of an effector, or by covalent modification such as phosphorylation. Ligand binding is normally rapidly reversible, and therefore the extent of binding and of allosteric regulation is due to mass action. The regulatory conformation is therefore always sensitive to the actual concentration of the regulatory ligand, and such control is very dynamic. In contrast, phosphorylation converts the enzyme into a different but stable conformation with a considerable lifetime, and this may be attained without accumulating high concentrations of a proximal effector.

The G proteins then show a third regulatory mode. While it is achieved by ligand binding rather than covalent modification, an affinity for the effector near nanomolar provides an extended lifetime of many minutes to hours for the bound state. The G protein, due to its ability to slowly alter the ligand, can return to the native state, but the kinetics provide for an extended period of activation in response to an external signal.

12.2 GEFs Lower the Affinity of p21 for Guanine Nucleotides

Because there are different types of G proteins, there also exists a variety of different GEFs specific for each type of GTP. The binding of a GEF to its cognate G protein leads to a conformational change in the G protein that lowers the binding affinity of GDP.

12.2.1 Changes in Affinity for Guanine Nucleotides Produced by GEF

The GEF protein also binds very tightly to another G protein, p21*ran* (Table 12.2). To make binding studies more efficient, GDP and GTP had been chemically modified by addition of a methyl-anthranylyl moiety that has a different spectroscopic emission in the free and bound states. There was only a minor change in the affinity of the modified nucleotides to p21,[288] with an affinity, K_G, in the picomolar range similar to that shown for the nucleotides, K_N. With p21*ran* GDP binds at about 6 pM, and the binding of GTP

Table 12.2. GEFs alter the affinity of p21*ran* for GTP and GDP[a]

Binding constant[b]	mGDP (M)	mGTP (M)
K_N	6.2×10^{-12}	1.4×10^{-10}
K_G	2.5×10^{-12}	2.5×10^{-12}
K_G^N	7.1×10^{-7}	5.3×10^{-7}
K_N^G	1.9×10^{-6}	2.9×10^{-5}

[a]mGDP is the methylanthraniloyl derivative of GDP. N, nucleotide; G,= GEF.
From Goody and Hofmann-Goody[288]
[b]For meaning of binding constants see Fig. 12.2

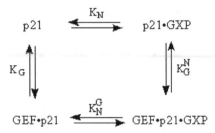

Fig. 12.2. Scheme for the reversible binding of guanine nucleotides (GXP) or exchange factors (GEF) to p21*ras*. Specific dissociation constants are shown

is also very strong at 140 pM. The principal effect of the conformational change produced by GEF binding to p21*ran* is that the affinity for both nucleotides, K_N^G, is weakened by five orders of magnitude (Table 12.2). Since GEF·p21·GDP has a K_d of 0.7 μM for GDP, this nucleotide is now able to dissociate readily. Although the GEF·p21*ras* complex has the same affinity for GTP as for GDP (Table 12.2), a very extensive exchange of GTP for GDP must occur since the cellular concentrations of GTP are generally about tenfold higher than for GDP.[31]

It should be noted that guanine nucleotides and GEF each change the affinity of p21 for binding the other ligand by a comparable extent (Table 12.2). This results from the thermodynamics of a reversible binding scheme for the three types of molecules that interact to form a binary and ternary complex (Fig. 12.2). For such a system

$$K_N K_G^N = K_G K_N^G. \tag{12.1}$$

This can be converted to:

$$\frac{K_N^G}{K_N} = \frac{K_G^N}{K_G}. \tag{12.2}$$

Since the direct affinities of either guanine nucleotides or GEF, when the other ligand is absent, are independent factors, the equality shown in (12.2) requires that the influence of GEF on the binding of guanine nucleotides must be matched by the influence of the nucleotides on the binding of GEF.

Table 12.3. Affinities of p21*ras* for guanine nucleotides

Ligand	K_d	Refs.
Guanosine	153 μM	289
GMP	29 μM	289
GDP	22 pM	288
GTP	0.5 nM	289

12.2.2 Specificity of the Guanine Nucleotide Binding Site

A remarkable feature of G proteins such as p21*ras* is the difference in affinity as the ligand guanosine acquires three additional phosphates. The nucleoside guanosine itself has a very modest affinity (Table 12.3). Adding the first phosphate, GMP, leads to a very modest increase in binding. Adding the second phosphate, GDP, improves binding by seven orders of magnitude. Adding the third phosphate must create some strain, since the binding is weaker by about 20-fold. Nevertheless, the contrast in affinities for GDP and GTP, relative to GMP, shows that the protein is designed to bind GDP while still providing a strong enough affinity for GTP that the latter can occupy the protein and stabilize the active conformation for an extended time.

This design permits the protein to be in the active conformation for a significant, though limited time, and after the hydrolysis of GTP, the very high affinity for GDP maintains the protein in the inactive form for an extended time. Only activation by external factors such as GEF can return the G protein to the active conformation.

12.2.3 Performing Binding Measurements for Tight Binding Ligands

Obtaining accurate binding values becomes a challenge when ligands bind so tightly that the purified protein will normally be almost fully bound with this ligand. If such a protein sample is not pretreated to produce free protein, then a binding experiment can only measure the extent to which exchange occurs between the bound ligand and the free experimental ligand being tested. This was made evident in efforts to measure the binding of radioactive GDP to newly purified p21*ras*. The authors observed that when the concentration of p21 was 8 nM, the apparent K_d was 7.3 nM, and when the concentration of p21 was 248 nM, the apparent K_d was 143 nM.[288] The fact that binding became weaker as the protein concentration was increased suggested that when the p21 protein sample was added to the binding assay, some of the bound GDP could slowly dissociate and thereby dilute the low concentration of the radioactive GDP. This would then require higher radioactive GDP to achieve binding and lead to the observed increase in the K_d at higher protein concentration.

In the present example the authors prepared ligand-free protein by incubating p21 with a phosphatase for several hours, and then added GPP-CH$_2$-P for the binding assay, since this analog of GTP is not cleaved by the phosphatase. This experimental design provided the binding constants shown in Table 12.3.

Table 12.4. Kinetics and oncogenic activity of mouse p21 mutants[a]

Enzyme	K_{obs} (s^{-1})	Oncogenic activity
Wild type	4.7×10^{-4}	–
G12V	3.3×10^{-5}	+
G12R	2.3×10^{-5}	+
Q61L	2.2×10^{-5}	+

[a]From Fech et al.[290]

Table 12.5. Kinetics with the human p21*ras*[a]

Protein	k_{diss}^{GDP} (s^{-1})	k_{diss}^{GTP} (s^{-1})	GTPase (s^{-1})
Wild type	1.3×10^{-4}	8.3×10^{-5}	4.7×10^{-4}
G13V	1.1×10^{-2}	5×10^{-4}	2.2×10^{-4}
E62H	1.5×10^{-4}	8.3×10^{-5}	1.8×10^{-4}

[a]Data from Gideon et al.[282]

12.3 Oncogenic Mutants of p21

G proteins may also serve as initiation and elongation factors at times. In this capacity they become oncogenic agents if they do not recycle to the inactive GDP state in the normal time frame. Numerous mutants of p21*ras* have been observed in rodents and humans.[287] For those that have the mutation at part of the catalytic site, it has been shown that the GTPase rate need only be decreased about 15–20-fold for transformation and oncogenesis to occur (Table 12.4). The extent of this decrease in the GTPase rate would lead to an equal increase in the lifetime of the active p21·GTP, during which the genes for enzymes and signals for cell division would become increasingly transcribed.

For the human p21*ras* the normal enzyme has the same GTPase activity as observed with the mouse enzyme (Tables 12.4 and 12.5). Human oncogenic variants are due to similar mutations and have similar effects on lowering the off rate for GTP. The lowering of the GTPase activity is not as great with the human mutants, being only about twofold, but this increased lifetime for the p21·GTP is enough to influence gene transcription leading to tumor development.

13

PROTEIN KINASES

Summary

Protein kinases are important in many aspects of cell signaling, cell maturation and differentiation, and metabolic control. Many of these enzymes are themselves regulated by phosphorylation, as well as by separate regulatory ligands. The effects of phosphorylation and ligand binding may also be synergistic. Many of these protein kinase activation signals are propagated in sequential protein phosphorylation steps, and the total activation can be very extensive for the final target enzyme. cAMP-dependent protein kinase provides an excellent example of how two ligands may each have a positive influence for the binding of the other one, even though the ligands bind to completely different subunits that are not in contact.

13.1 Overview of Protein Kinases

Phosphorylation of enzymes to stabilize an alternate conformation may be the single most widely applied regulatory mechanism. Based on sequence comparisons to the 493 known protein kinases in *C. elegans*, it is estimated that more than 1,100 human genes code for such enzymes.[291] Many of these protein kinases belong to subsets of related members, of which the larger groups are listed in Table 13.1. For many of these the scientific effort has largely been on determining the diversity of isogenes and the cellular response patterns, and structural information and reliable kinetic data are not available. Our discussion in this chapter will focus on those that have been sufficiently well characterized.

The complexity presented by the overlapping patterns for the activation and function of even a single protein kinase family is illustrated by the MAP kinases (Fig. 13.1). A variety of external signals may interact with specific membrane receptors, and these in turn activate an intracellular G protein, or protein kinase (not included in Fig. 13.1). The G protein or protein kinase then leads to the activation of a MAPKKK enzyme, either by

Table 13.1. Protein kinase families

Enzyme name	Definition	Function	Structure[a]
AMPK	AMP activated protein kinase	Metabolic sensor; fuel gauge	$\alpha\beta\gamma$
cAPK, (PKA)	cAMP-dependent protein kinase		C_2R_2
MAPK	Mitogen activated protein kinase	Alter gene expression for differentiation. growth, immunity	α_2
PKA	Protein kinase A, cAMP dependent		C_2R_2

[a]C, catalytic subunit; R, regulatory subunit

binding (G protein) or by the phosphorylation of the MAPKKK[*] (by the initial protein kinase). It is a general feature of protein kinases that they have these dual mechanisms for becoming activated, since almost all of them bind an activating ligand, and are also themselves targets for phosphorylation. Note that each initial signal leads to the activation of more than one MAPKKK, and each MAPKKK is activated by more than one initial signal (Fig. 13.1).

The activated MAPKKK enzymes in turn phosphorylate their designated targets (MAPKKs), and these then phosphorylate the principal MAPK enzymes. Activation of these MAPK enzymes leads to the activation of diverse subsets of target proteins, to accomplish a variety of physiological results (Fig. 13.1). Some refinement of this apparently overlapping diversity is obtained by the limited or selective expression of each MAPKKK and MAPKK and MAPK enzyme in appropriate cells or tissues, as exemplified for the p38 protein kinase isoenzymes (Table 13.2).

A very significant aspect of such a cascade is that each successive round of phosphorylation on downstream targets amplifies the initial stimulus. This was briefly described in Chap. 4.1.3 for normal metabolic enzymes in glycogen metabolism.

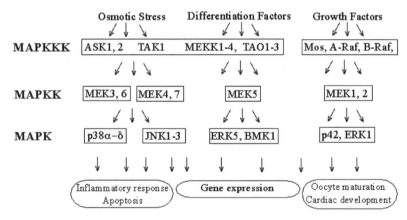

Fig. 13.1. Sequential activation of mitogen-activated protein kinases. Physiological changes or external signals may lead to the activation of specific protein kinases, with amplification of this signal by three sequential MAP kinases

---

[*]A MAPKKK is a mitogen-activated protein kinase kinase kinase since it phosphorylates a MAPKK which in turn phosphorylates a MAPK. Some writers use MAP3K and MAP2K.

Table 13.2. Isoform expression of human p38 MAP kinases

Isoenzyme	Cellular expression	k_{cat} (s^{-1})	K_m (μM)^a	k_{cat}/K_m (s^{-1}M^{-1})	Refs.
p38α	All cells	0.4	170	2.4×10^3	292
p38β	T cells, endothelial cells	0.001	9.2	1.1×10^2	293
p38δ	T cells, macrophages, neutrophils				
p38γ	Nonimmume cells	1.04	13	8.0×10^4	292

^{a}Affinity for ATP

The highly desirable result of such a multistep cascade is that a modest change in the concentration of some initial signal molecule is amplified increasingly with each successive step. Therefore, all cascade systems as a set of enzymes must belong to the V-type family for allosteric regulation.

13.2 Map Kinases

MAP kinases were originally named for their activation by mitogens or cellular growth factors. The family has become quite large and Fig. 13.1 shows a diversity of original MAPKKK enzymes and a representative set of the downstream MAPK enzymes.[294] Because they have been so well characterized, we will consider two MAP kinases, p38 and p42.

13.2.1 p38 MAP Kinase

Four isoenzymes for human p38 MAP kinase are currently known. Consistent with their roles as regulatory switches, they normally have very modest enzyme activities (Table 13.2). Nevertheless, while the actual structure to account for this has not yet been determined, it is worth noting that p38γ has a thousandfold greater activity than p38β. Since p38γ is the only enzyme that has an almost normal specificity constant, it may not need to be activated to the same extent as the other isoenzymes.

A more expanded scheme for the activation of this single MAP kinase shows the cellular receptors plus the internal kinases that are involved (Fig. 13.2). Three membrane receptors are known to bind extracellular stimulus molecules, such as interleukins (IL-12R), immunoglobulins (CD28), and T cells (TCR). The intracellular domain of such receptors, upon being activated, stimulates their immediate target enzyme. For two of the receptors this is the GADD kinase (for growth arrest/DNA damage). The TCR receptor interacts directly with p38, leading to phosphorylation of Tyrosine 323. Phosphorylation results in a conformational change that helps to promote the dual phosphorylation of the two required residues, Threonine 180 and Tyrosine 182.

Activation of p38 requires that it be phosphorylated at these two separate residues, Threonine 180 and Tyrosine 182.[295] This was verified by using a T180A mutant, which no longer has the Threonine 180 phosphorylation site. The mutant with only one phosphorylated site was catalytically inactive. The extent of activation of p38 varies with the original stimulus that initiates the MAP kinase cascade, as well as with the time after the signal is received. The activation pattern for multiple stimuli is shown in Fig. 13.3.

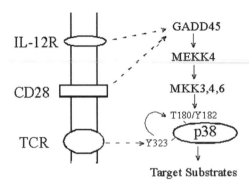

Fig. 13.2. The activation cascade for p38 in T cells. Membranes contain the T cell receptor (TCR) as well as the receptors for CD28 and interleukin. These act on the growth arrest/DNA damage (GADD) inducible genes, which then activate the MAP kinases MEKK4 etc. leading to phosphorylation of p38 at two residues

Human cells in culture were exposed to the appropriate stimulus for the time shown, at which cells were extracted and the p38 immunoprecipitated, so that it could later be tested for kinase activity. By employing radioactive ATP the resolution for this assay was sufficient to detect even low levels of enzyme activity.

The extent of activation varied from about twofold with interleukin or epidermal growth factor (EGF), up to 16-fold with UV radiation. The time curves show that optimum activation is normally achieved by 30 min, though for osmotic shock this lasted longer. The nature of the activity assay could not distinguish whether only the protein present at time zero was activated, or whether new protein had been synthesized and activated. However, the time curve shows that large increases in activity are occurring faster than could occur by gene expression. Since the p38 activity being monitored results from several sequential protein phosphorylation steps, the time is very consistent with a process whereby mostly existing p38 is being activated.

The most dramatic activation occurs as a consequence of UV radiation, and must reflect that such radiation very effectively activates the immediate MAPKKK. Tumor necrosis factor (TNF) is also a very good agonist. Other stimuli that are known to influence some MAP kinases, EGF, interleukin-1 (ILK1), and phorbol ester, had only a very modest degree of activation of p38. This shows that while various members in the different cascades (Fig. 13.1) may cross-react, they are still somewhat specific for particular downstream targets.

13.2.1.1 Inhibition of p38 at a Novel Allosteric Site

Cryptic allosteric sites were described in Chap. 4.2.2 for a number of different enzymes. These sites were thought of as cryptic since they were separate and distinct from the sites known to bind the substrates. A similar discovery has occurred with p38, but with the additional feature that the site exists only in one distinct conformation of the enzyme. Near the ATP site on the enzyme there exists a conserved triplet of amino acids Aspartate-Phenylalanine-Glycine (DFG).[133]. The side chains of these three residues are normally positioned so as to be at the bottom of a cavity, facing toward the center of the

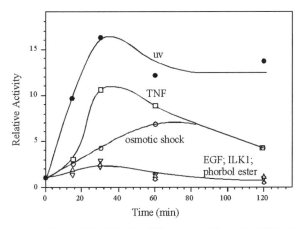

Fig. 13.3. The extent of activation of p38 MAPK after different stimuli, including: UV radiation, tumor necrosis factor (TNF), epidermal growth factor (EGF), interleukin-1 (ILK1). Data from Raingeaud et al.[295]

enzyme core. A less stable conformation of the enzyme exists in which the DFG side chains flip outward, and this altered conformation prevents the binding of ATP.

Because p38 leads to an inflammatory response, this activity can be associated with various chronic inflammatory disorders when it is too active. This has led to a search for specific inhibitory drugs, and one of these (BIRB 796) has been shown to bind uniquely to the enzyme conformation with the DFG residues in the outward orientation. The drug has a K_d of 0.1 nM, and thus stabilizes the enzyme in this conformation that is normally at a low frequency. Since ATP cannot bind to this conformation, the drug is not a competitive inhibitor.[133]

13.2.2 p42 MAP Kinase

As shown in Fig. 13.1, stimulation by a growth factor activates the MAPKKK Mos, which activates MEK1, leading to activation of p42. The penultimate stimulus for activating cell division in *Xenopus* oocytes is the phosphorylated active form of the p42 MAPK.[296] For these studies freshly prepared extracts of *Xenopus* oocytes received the indicated addition of pure Mos enzyme, and p42 activity was measured after the system had approached a steady state.

The remarkable on/off switch produced by only a threefold change in the concentration of the initial MAPKKK, Mos, leads to an almost 20-fold change in the activity of p42 (Fig. 13.4). When the concentration of Mos is at 20 nM, the activity of p42 is less than 5% of the maximum possible activity. When Mos has increased from 20 to 60 nM, the activity of p42 is greater than 90% of the maximum. This large change in activity is described by a Hill coefficient of 4.9, an indicator of very high positive cooperativity. However, if the kinetics of p42 are evaluated directly, and not as part of a cascade, the enzyme has a classic hyperbolic activity curve, and shows no evidence of cooperativity.[296] This demonstrates the amplification effect inherent in any cascade with multiple enzymes, as was discussed in Chap. 5.5.1.

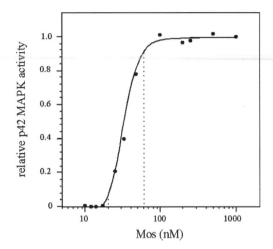

Fig. 13.4. The cellular increase in activity of p42 MAPK with an increase in activity of Mos MAPKKK. *Dashed lines* indicate concentrations of Mos at 20 and 60 nM. Data from Huang and Ferrel[296]

The great sensitivity or ultrasensitivity displayed in Fig. 13.4 is a property that has been described by Stadtman[149,150] by Koshland[297, 298] and by Ferrel[152] for a variety of enzyme cascades. These papers provide mathematical models, described in Chap. 5.5.1, that illustrate the sharp response by the final enzyme in a cascade to a small, but physiologically appropriate change in the concentration of the initial signal molecule. An interesting implication from such models is that a cascade should have at least three successive activating enzymes for an ultrasensitive response. This number is well supported by the many enzymatic cascades that have been defined, since three is the minimum number of enzymes normally observed.

13.2.3 AMP-Activated Protein Kinase

Because of its functions relative to energy metabolism (Table 13.3), AMP-activated protein kinase (AMPK) has been described as the fuel gauge for the cell.[299] Due to the importance of ATP as the most frequently used molecule for the many biosynthetic reactions in general metabolism, the cell has multiple mechanisms to sense the steady-state ATP concentration, and to respond to any changes with compensatory mechanisms. Should ATP not be synthesized as rapidly as it is being consumed, AMP concentrations will increase proportionately. This results from the action of adenylate kinase to convert

Table 13.3. Functions of AMP-activated protein kinase

Activation:	Glucose uptake	Glycolysis	Fatty acid oxidation	Mitochondrial biogenesis
Inhibition:	Gluconeo genesis	Glycogen synthesis	Fatty acid synthesis	Protein synthesis

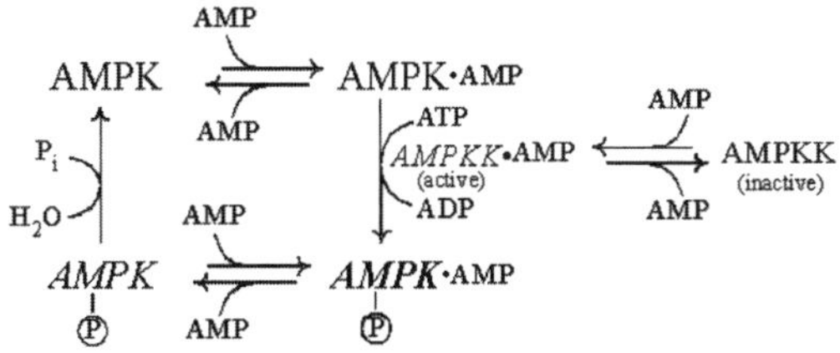

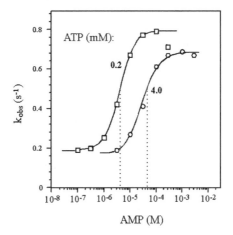

Fig. 13.5. Regulation of AMP-activated protein kinase (AMPK) by binding AMP, and/or by phosphorylation. The P-AMPK·AMP enzyme (*bold* in the figure) has the highest activity

ADP back to ATP (2 ADP $\rightleftarrows$ ATP + AMP). To be a sensitive regulator the responsible enzyme should therefore be inhibited by ATP and also be activated by AMP. These are the regulatory features that make AMPK an important regulatory enzyme.

The enzyme is activated by both ligand binding with AMP, and by phosphorylation by a MAPKK, and therefore multiple species may exist in a steady state mixtures as shown (Fig. 13.5). The most active species is phosphorylated and also bound with AMP. The enzyme appears to have a very modest activity, both in the unmodified basal state, and even after being activated (Fig. 13.6). It is evident that at saturating AMP concentrations the change in activity is only about 3.5–4-fold. This same degree of activation has been found for this enzyme from other sources: rat skeletal muscle[300] and bovine heart.[301] This may well be the natural limit for the increase in activity possible with the conformation stabilized by AMP.

The enzyme form that is most active is the phosphorylated form that is also bound with AMP. This was demonstrated with the enzyme purified from rat liver. Part of this enzyme preparation was subjected to phosphorylation by the purified MAPKK that

Fig. 13.6. Activation of AMP-activated protein kinase from rat hepatocytes by AMP, and inhibition by ATP. Data from Corton et al.[302]

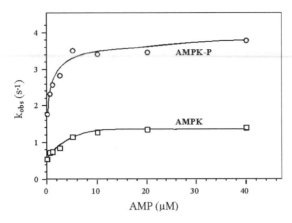

Fig. 13.7. The effect of enzyme phosphorylation on the activation of AMPK by AMP. Data from Weekes et al.[303]

phosphorylates AMPK. Both forms of AMPK were then tested for increased enzyme activity when incubated with AMP (Fig. 13.7). As described above, the native enzyme demonstrated only a threefold change in activity above the basal level of activity in the absence of AMP. By comparison, the phosphorylated enzyme, AMPK-P, already had greater than threefold the activity of the native enzyme, and then had a further doubling in activity upon binding with AMP. When all these results are compared (Table 13.4), it appears that a change of up to tenfold in V_{max} may be possible with both types of allosteric modification.

Crystal structures for AMPK from yeast have been obtained with both AMP and ATP bound.[304] Surprisingly these structures shed no light on how the regulatory mechanism is achieved. Both AMP and ATP bind to a single site on the γ subunit. They make almost identical contacts with protein side chains, and there is no observable difference in the overall structure of the enzyme. Since neither ATP nor AMP were observed at the catalytic site of the α subunit in these crystals, it is possible that some aspect of the crystallization conditions favored the inhibited conformation so that the catalytic site would remain empty while the regulatory site accommodated both the activator and the inhibitor nucleotide.

Table 13.4. Extent of activation of AMPK

Enzyme form	Hepatocyte enzyme[a] k_{cat} (s^{-1})	Rat liver enzyme[b] k_{cat} (s^{-1})
AMPK	0.2	0.5
AMPK·AMP	0.8	1.4
AMPK-P		1.7
AMPK-P·AMP		3.8

[a]From Corton et al.[302]
[b]From Weekes et al.[303]

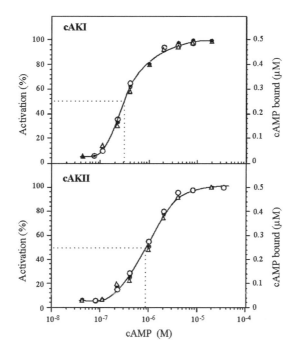

Fig. 13.8. Cooperative binding of cAMP to the two protein kinase isoenzymes from bovine heart. (*Filled circle*) Activation of the enzyme by cAMP. Direct binding to site A (*open circle*) and site B (*open triangle*). Data from Øgreid and Døskeland[305]

13.2.4 cAMP-Activated Protein Kinase

Because of their importance in the regulation of glycogen metabolism, cAMP-dependent protein kinases (cAPKs) were among the first protein kinases to be discovered.[306] These ubiquitous enzymes have the structure C_2R_2, in which the two catalytic subunits are stabilized in the inactive conformation via binding to the regulatory dimer.[307] Binding of cAMP to the R subunits leads to dissociation and formation of the active catalytic monomer: $C_2R_2 + 4$ cAMP $\rightleftarrows R_2 \cdot cAMP_4 + 2$ C. This binding scheme reflects the fact that each regulatory subunit contains two sequential sites for the binding of cAMP (site A and site B).

There is interaction between the two cAMP sites on a single subunit, as shown by binding curves. Since many species have isozymes of cAMP-dependent kinase, the two isozymes from bovine muscle were tested for their extent of binding as well as their extent of activation as a function of cAMP concentration (Fig. 13.8).[305] Several interesting features are evident in these two figures. Binding of cAMP to the R subunits occurs with positive cooperativity. It is possible to distinguish binding to the A site, and since one can easily measure total binding, then binding to the B site is the difference. There is no evident difference in binding of cAMP to the two sites. Furthermore, the degree of enzyme activation also shows the identical binding function.

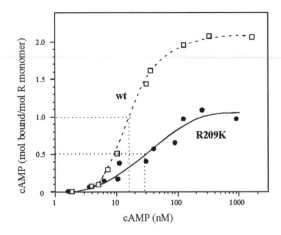

Fig. 13.9. A mutant subunit for the enzyme from pig heart with one functional cAMP binding site. Data from Bubis et al.[308]

While the two cAMP sites on one R subunit appear to be the same, there is about a threefold difference in the K_d for cAMP between the two isozymes. It is not yet clear if the separate isozymes have different downstream targets. If they have the same or at least overlapping targets, then the different affinities observed (Fig. 13.8) would function similar to negative cooperativity to provide a greater concentration range for stimulating a cellular response.

13.2.4.1 Both cAMP Sites are Not Required for Activation

By deducing the conserved amino acid residue necessary for binding the phosphate of cAMP, it was possible to make a single site mutant that no longer binds cAMP at the A site.[308] With a clever replacement of the conserved arginine with a lysine in the enzyme from pig heart (R209K mutation), the protein structure maintains the same positive charge at that position, but no longer has enough room to accommodate the ligand, cAMP. This mutant now binds only one mole cAMP per mole R subunit, compared to the native enzyme that binds two moles per R subunit (Fig. 13.9).

This figure shows strong positive cooperativity for the native enzyme from pig muscle. The mutant has a lowered n_H of 1.13, indicating that there is almost no cooperativity. Since the regulatory dimer (R_2) has four cAMP sites, the experiment with the mutant shows that the cooperativity is largely between sites in the same protein subunit. The mutant also demonstrates that binding at the remaining B site is modestly affected by the A site mutation, since the single B site now has a threefold higher $K_{0.5}$ for binding cAMP. In separate experiments it was demonstrated that the mutant enzyme had almost the same activation response to cAMP, thereby proving that binding at both sites is not required for the enzyme to become activated.

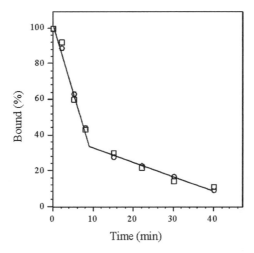

Fig. 13.10. Binding of cAMP to the regulatory subunits of cAMP-dependent protein kinase from bovine heart: isozyme RI (*open square*) and RII (*open circle*). Equilibrium was first established with ^{3}H-cAMP at 1 µM; then 1 mM unlabeled cAMP was added, and the dissociation of radioactive cAMP was measured. Data from Rannels and Corbin[309]

13.2.4.2 The Two cAMP Sites Have Different Dissociation Rates

To measure the difference in the K_d for cAMP between sites A and B, the dissociation of saturating cAMP was monitored, since the off rate is directly indicative of the affinity (Fig. 13.10). The enzyme was first equilibrated with saturating radioactive cAMP. The preparation was then filtered, and an excess of unlabeled cAMP (1 mM) was added. At intervals samples were measured for the remaining radioactivity bound. The results show a clear biphasic dissociation curve, indicating a tenfold difference in the binding at the two sites. The results were identical for the two isoenzymes found in bovine heart muscle. Although visually the break in the dissociation curve occurs when more than 60% of the ligand has been released, when the data for the fast dissociation are corrected for the contribution from the slow dissociation, then each component contributes $50 \pm 5\%$ to the result.[309]

13.2.4.3 ATP Alters the Affinity for cAMP

Since cAPK is a kinase it must bind ATP at the catalytic site, and such binding is greatly improved by the binding of cAMP to the R subunits, since this leads to dissociation of the R_2 dimer from the heterotetramer. We can view this as a steady-state system where R dimers are stabilized by cAMP and C monomers are stabilized by ATP (13.1). If ATP and cAMP each bind to the free C or R proteins (13.1) then the formation of the inactive R_2C_2 tetramer is diminished by the presence of either ligand. The actual influence of ATP on the binding of cAMP was tested with cAPKII from bovine heart (Fig. 13.9).[305] It is evident in this experiment that the binding of ATP at the C subunit also improves the affinity for the binding of cAMP to the R subunit.

$$
\begin{array}{c}
\text{C·ATP} \\
\text{ATP} \uparrow\downarrow \\
R_2 + 2C \rightleftarrows R_2C_2 \\
\downarrow\uparrow \text{cAMP} \\
R_2\cdot\text{cAMP}_4
\end{array}
$$

(13.1)

Note that, as diagrammed in (13.1), each of the two different ligands influences the binding of the other ligand even though they are binding to enzyme subunits that are not in physical contact with each other. This experiment therefore provides an excellent illustration that the meaning of "induced fit" should be interpreted in terms of how the ligand influences the distribution of enzyme conformations, either at equilibrium or in a steady state system. As diagrammed (13.1) neither ligand has any significant interaction with the native free cAPK tetramer, which is the inactive form of this enzyme.

These results should also be seen as very different from the competition between ATP and AMP for the AMP-activated protein kinase (Fig. 13.6). In that experiment ATP and AMP both bind to the same regulatory site, so that competition must occur. With the cAPK enzyme, ATP and cAMP bind to completely separate sites on different subunits, making a positive effect possible (Fig. 13.11).

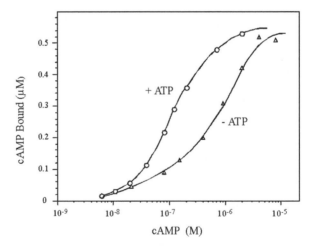

Fig. 13.11. Cooperativity for the binding of cAMP to cAMP-dependent protein kinase from bovine heart. Total binding for cAMP to sites A and B, (*open triangle*) without and (*open circle*) with 3 mM ATP. Data from Øgreid and Døskeland[305]

REFERENCES

1. Y. Ogura, Catalase activity at high concentration of hydrogen peroxide. *Arch. Biochem. Biophys.* **57**, 288–300 (1955).
2. H. Steiner, B.-H. Jonsson and S. Lindskog. The catalytic mechanism of carbonic anhydrase. *Eur. J. Biochem.* **59**, 253–259 (1975).
3. J. N. Earnhardt, M. Qian, C. Tu, M. M. Lakkis, N. C. H. Bergenhem, P. J. Laipis, R. E. Tashian and D. N. Silverman, The catalytic properties of murine carbonic anhydrase VII. *Biochemistry* **37**, 10837–10845 (1998).
4. A. Radzicka and R. Wolfenden, A proficient enzyme. *Science* **267**, 90–93 (1995).
5. D. Herschlag and T. R. Cech, Catalysis of RNA cleavage by the *Tetrahymena thermophila* ribozyme. *Biochemistry* **29**, 10159–10171 (1990).
6. D. M. J. Lilley, The origin of RNA catalysts. *Trends Biochem. Sci.* **28**, 495–501 (2003).
7. S. W. Santoro and G. F. Joyce, A general purpose RNA-cleaving DNA enzyme. *Proc. Natl Acad. Sci. USA* **94**, 4262–4266 (1997).
8. D. N. Bolon, De novo design of biocatalysts. *Curr. Opin. Chem. Biol.* **6**, 125–129 (2002).
9. R. Wolfenden and M. J. Snider, The depth of chemical time and the power of enzymes as catalysts. *Acc. Chem. Res.* **34**, 938–945 (2001).
10. B. G. Miller, A. M. Hassell, R. Wolfenden, M. V. Milburn and S. A. Short, Anatomy of a proficient enzyme: The structure of orotidine 5'-phosphate decraboxylase in the presence and absence of a potential transition state analog. *Proc. Natl Acad. Sci. USA* **97**, 2011–2016 (2000).
11. Y. Bai, T. R. Sosnick, L. Mayne and S. W. Englander, Protein folding intermediates: native-state hydrogen exchange. *Science* **269**, 192–197 (1995).
12. S. Kumar, B. Ma, C.-J. Tsai, N. Sinha and R. Nussinov, Folding and binding cascades: dynamic landscapes and population shifts. *Protein Sci.* **9**, 10–19 (2000).
13. M. Lynch and J. S. Conery, The evolutionary fate and consequences of duplicate genes. *Science* **290**, 1151–1155 (2000).
14. R. J. Britten, The majority of human genes have regions repeated in other genes. *Proc. Natl Acad. Sci. USA* **102**, 5466–5470 (2005).
15. A. J. Fulco, Fatty acid metabolism in bacteria. *Prog. Lipid Res.* **22**, 133–160 (1983).
16. J. K. Stoops and S. J. Wakil, Yeast fatty acid synthetase: Structure–function relationship and nature of the β-ketoacyl synthetase site. *Proc. Natl Acad. Sci. USA* **77**, 4544–4548 (1980).
17. Y. Tsukamoto, H. Wong, J. S. Mattick and S. J. Wakil, The architecture of the animal fatty acid synthetase complex. IV. Mapping of active centers and model for the mechanism of action. *J. Biol. Chem.* **258**, 15313–15322 (1983).
18. H. Ardehali, Y. Yano, R. L. Printz, S. Koch, R. R. Whitesell, J. M. May and D. K. Granner, Functional organization of mammalian hexokinase II. *J. Biol. Chem.* **271**, 1849–1852 (1996).
19. H. Nyunoya and C. Lusty, The *carB* gene of *Escherichia coli*: A duplicated gene coding for the large subunit of carbamoyl-phosphate synthetase. *Proc. Natl Acad. Sci. USA* **80**, 4629–4633 (1983).
20. X. Liu, C. S. Kim, F. T. Kurbanov, R. B. Honzatko and H. J. Fromm, Dual mechanisms for glucose 6-phosphate inhibition of human brain hexokinase. *J. Biol. Chem.* **274**, 31155–31159 (1999).
21. R. A. Poorman, A. Randolph, R. G. Kemp and R. L. Heinrikson, Evolution of phosphofructokinase–gene duplication and creation of new effector sites. *Nature* **309**, 467–469 (1984).

22. J. R. H. Tame, K. Namba and E. J. Dodson, The crystal structure of HpcE, a bifunctional decarboxylase/isomerase with a multifunctional fold. *Biochemistry* **41**, 2982–2989 (2002).

23. J. Monod, J. Wyman and J.-P. Changeux, On the nature of allosteric transitions: a plausible model. *J. Mol. Biol.* **12**, 88–118 (1965).

24. D. E. Koshland Jr., G. Némethy and D. Filmer, Comparison of experimental binding data and theoretical models in proteins containing subunits. *Biochemistry* **5**, 365–385 (1966).

25. T. W. Traut, Dissociation of enzyme oligomers: a mechanism for allosteric regulation. *CRC Crit. Rev. Biochem. Mol. Biol.* **29**, 125–163 (1994).

26. Y. Liu and D. Eisenberg, 3D domain swapping: as domains continue to swap. *Protein Sci.* **11**, 1285–1299 (2002).

27. R. Piccoli, A. Di Donato and G. D'Alessio, Co-operativity in seminal ribonuclease function. *Biochem. J.* **253**, 329–336 (1988).

28. E. K. Jaffe, Morpheeins – a new structural paradigm for allosteric regulation. *Trends Biochem. Sci.* **30**, 490–497 (2005).

29. A. B. Fulton, How crowded is the cytoplasm? *Cell* **30**, 345–347 (1982).

30. T. Masuda, G. P. Dobson and R. L. Veech, The Gibbs–Donnan near-equilibrium system of heart. *J. Biol. Chem.* **265**, 20321–20334 (1990).

31. T. W. Traut, Physiological concentrations of purines and pyrimidines. *Mol. Cell. Biochem.* **140**, 1–22 (1994).

32. Y. Kashiwaya, K. Sato, N. Tsuchiya, S. Thomas, D. A. Fell, R. L. Veech and J. V. Passonneau, Control of glucose utilization in working perfused rat heart. *J. Biol. Chem.* **269**, 25502–25514 (1994).

33. BRENDA. http://www.brenda.uni-koeln.de/.

34. P. Srere and H. R. Knull, Location – location – location. *Trends Biochem. Sci.* **23**, 319–320 (1998).

35. G. Wistow and J. Piatigorsky, Recruitment of enzymes as lens structural proteins. *Science* **236**, 1554–1556 (1987).

36. M. J. Yablonski, D. A. Pasek, B.-D. Han, M. E. Jones and T. W. Traut, Intrinsic activity and stability of bifunctional human UMP synthase and its two separate catalytic domains, orotate phosphoribosyltransferase and orotidine-5'-phosphate decarboxylase. *J. Biol. Chem.* **271**, 10704–10708 (1996).

37. C. P. Whitman, B. A. Aird, W. R. Gillespie and N. J. Stolowich, Chemical and enzymatic ketonization of 2-hydroxymuconate, a conjugated enol. *J. Am. Chem. Soc.* **113**, 3154–3162 (1991).

38. R. M. Pollack, B. Zeng, J. P. G. Mack and S. Eldin, Determination of the microscopic rate constants for the base-catalyzed conjugation of 5-androstene-3,17-dione. *J. Am. Chem. Soc.* **111**, 6419–6423 (1989).

39. W. Kuhne, Ueber das Verhalten verschiedener organisirter und sog. ungeformter Fermente. *Verh. Heidel. Natur.-Med. Ver.* **1**, 194–199 (1876).

40. E. Buchner, Alkoholische Gährung ohne Hefezellen. *Ber. Dtsch. Chem. Gesell.* **30**, 117–124 (1897).

41. Payen and Persoz, Mémoire sur la diastase, les principaux produits de ses réactions, et leurs applications aux arts industriels. *Ann. Chim. Phys.* **53**, 73–92 (1833).

42. W. F. Ostwald, Ueber die Katalyse. *Z. physikal. Chem.* **15**, 705–706 (1894).

43. A. J. Brown, Enzyme action. *J. Chem. Soc.* **81**, 373–388 (1902).

44. V. Henri, Théorie générale de l'action de quelques diastases. *C. R. Séances Acad. Sci.* **135**, 916–919 (1902).

45. L. Michaelis and M. L. Menten, Die Kinetik der Invertinwirkung. *Biochem. Z.* **49**, 333–369 (1913).

46. J. B. Sumner, The isolation and crystallization of the enzyme urease. *J. Biol. Chem.* **69**, 435–441 (1926).

47. D. E. Koshland Jr., Applications of a theory of enzyme specificity to protein synthesis. *Proc. Natl Acad. Sci. USA* **44**, 98–104 (1958).

48. J. C. Gerhart and A. B. Pardee, The enzymology of control by feedback inhibition. *J. Biol. Chem.* **237**, 891–896 (1962).

49. J. Monod, J.-P. Changeux and F. Jacob, Allosteric proteins and cellular control systems. *J. Mol. Biol.* **6**, 306–329 (1963).

50. C. Tanford and J. Reynolds, *Nature's Robots*. (Oxford University Press, Oxford, 2001).

51. J. A. Hathaway and D. E. Atkinson, The effect of adenylic acid on yeast nicotinamide adenine dinucleotide isocitrate dehydrogenase, a possible metabolic control mechanism. *J. Biol. Chem.* **238**, 2875–2881 (1963).

52. R. Okazaki and A. Kornberg, Deoxythymidine kinase of *Escherichia coli* II. Kinetics and feedback control. *J. Biol. Chem.* **239**, 275–284 (1964).

53. H. E. Kubitschek and J. A. Friske, Determination of bacterial cell volume with the Coulter Counter. *J. Bacteriol.* **168**, 1466–1467 (1986).

54. J. B. Stock, B. Rauch and S. Roseman, Periplasmic space in *Salmonella typhimurium* and *Escherichia coli*. *J. Biol. Chem.* **252**, 7850–7861 (1977).

55. J. A. Imlay and I. Fridovich, Assay of metabolic superoxide production in *Escherichia coli*. *J. Biol. Chem.* **266**, 6957–6965 (1991).

56. A. S. Hearn, L. Fan, J. R. Lepock, J. P. Luba, W. B. Greenleaf, D. E. Cabelli, J. T. Tainer, H. S. Nick and D. N. Silverman, Amino acid substitution at the dimeric interface of human manganese superoxide dismutase. *J. Biol. Chem.* **279**, 5861–5866 (2004).

57. A. S. Relman, E. J. Lennon and J. Lemann Jr., Endogenous production of fixed acid and the measurement of the net balance of acid in normal subjects. *J. Clin. Invest.* **40**, 1621–1630 (1961).

58. I. H. Kavakli and A. Sancar, Analysis of the role of intraprotein electron transfer in photoreactivation by DNA photolyase *in vivo*. *Biochemistry* **43**, 15103–15110 (2004).

59. I. Fridovich, Superoxide radicals: an endogenous toxicant. *Annu. Rev. Pharmacol. Toxicol.* **23**, 239–257 (1983).

60. A. Meisel, P. Mackeldanz, T. A. Bickle, D. H. Krüger and C. Schroeder, Tyoe III restriction endonucleases translocate DNA in a reaction driven by recognition site-specific ATP hydrolysis. *EMBO J.* **14**, 2958–2966 (1995).

61. A. Pingoud and A. Jeltsch, Recognition of DNA by type-II restriction endonucleases. *Eur. J. Biochem.* **246**, 1–22 (1997).

62. V. Pingoud, A. Sudina, H. Geyer, J. M. Bujnicki, R. Lurz, G. Lüder, R. Morgan, E. Kubareva and A. Pingoud, Specificity changes in the evolution of type II restriction endonucleases. *J. Biol. Chem.* **280**, 42890–4298 (2005).

63. P. Janscak, A. Abadjieva and K. Firman, The type I Restriction endonuclease R.*Eco*R124I: Overproduction and biochemical properties. *J. Mol. Biol.* **257**, 977–991 (1996).

64. E. K. Jameson, R. A. Roof, M. R. Whorton, H. I. Mosberg, R. K. Sunahara, R. R. Neubig and R. T. Kennedy, Real-time detection of basal and stimulated G protein GTPase activity using fluorescence GTP analogues. *J. Biol. Chem.* **280**, 7712–7719 (2005).

65. M. Nakajima, K. Imai, H. Ito, T. Nishiwaki, Y. Murayama, H. Iwasaki, T. Oyama and T. Kondo, Reconstitution of circadian oscillation of cyanobacterial KaiC phosphorylation in vitro. *Science* **308**, 414–415 (2005).

66. P. Miller, A. M. Zhabotinsky, J. E. Lisman and X.-J. Wang, The stability of a stochastic CaMKII switch: dependence on the number of enzyme molecules and protein turnover. *Public Lib. Sci. Biol.* **3**, 0705–07016 (2005).

67. J. D. Petersen, X. Chen, L. Vinade, J. E. Lisman and T. S. Reese, Distribution of postsynaptic density (PSD)-95 and Ca^{2+}/calmodulin-dependent protein kinase II at the PSD. *J. Neurosci.* **23**, 11270–11278 (2003).

68. A. Fersht, *Structure and Mechanism in Protein Science*. (W. H. Freeman and Co., New York, 1999).

69. C. F. Hawkins and A. S. Bagnara, Adenosine kinase from human erythrocytes: kinetic studies and characterization of adenosine binding sites. *Biochemistry* **26**, 1982–1987 (1987).

70. R. L. Miller, D. L. Adamczyk and W. H. Miller, Adenosine kinase from rabbit liver. *J. Biol. Chem.* **254**, 2339–2345 (1979).

71. M. Johansson, A. R. van Rompay, B. Degrève, J. Balzarini and A. Karlsson, Cloning and characterization of the multisubstrate deoxyribonucleotide kinase of *Drosophila melanogaster*. *J. Biol. Chem.* **274**, 23814–23819 (1999).

72. R. Chakravarty, S. Ikeda and D. H. Ives, Distinct sites for deoxyguanosine and deoxyadenosine phosphorylation on a monomeric kinase from *Lactobacillus acidophilus*. *Biochemistry* **23**, 6235–6240 (1984).

73. I. Park and D. H. Ives, Properties of a highly purified mitochondrial deoxyguanosine kinase. *Arch. Biochem. Biophys.* **266**, 51–60 (1988).

74. R. C. Payne, N. Cheng and T. W. Traut, Regulation of uridine kinase. *J. Biol. Chem.* **261**, 13006–13012 (1986).

75. T. Okajima, K. Tanizawa and T. Fukui, Site-directed mutagenesis of AMP-binding residues in adenylate kinase. *FEBS Lett.* **334**, 86–88 (1993).

76. T. Dahnke and M.-D. Tsai, Mechanism of adenylate kinase. *J. Biol. Chem.* **269**, 8075–8081 (1994).

77. I. S. Girons, A.-M. Gilles, D. Margarita, S. Michelson, M. Monnot, S. Fermandjian, A. Danchin and O. Bârzu, Structural and catalytic characteristics of *Escherichia coli* adenylate kinase. *J. Biol. Chem.* **262**, 622–629 (1987).

78. J. Reinstein, M. Brune and A. Wittinghofer, Mutations in the nucleotide binding loop of adenylate kinase of *Escherichia coli. Biochemistry* 27, 4712–4720 (1988).

79. D. G. Rhoads and J. L. Lowenstein, Initial velocity and equilibrium kinetics of myokinase. *J. Biol. Chem.* 243, 3963–3972 (1968).

80. C. P. Schultz, L. Ylisastigui-Pons, L. Serina, H. Sakamoto, H. H. Mantsch, J. Neuhard, O. Bârzu and A.-M. Gilles, Structural and catalytic properties of CMP kinase from *Bacillus subtilis*: a comparative analysis with the homologous enzyme from *Escherichia coli. Arch. Biochem. Biophys.* 340, 144–153 (1997).

81. N. Bucurenzi, H. Sakamoto, P. Briozzo, N. Palibroda, L. Serina, R. S. Sarfati, G. Labesse, G. Briand, A. Danchin, O. Bârzu, and A.-M. Gilles, CMP kinase from *Escherichia coli* is structurally related to other nucleoside monophosphate kinases. *J. Biol. Chem.* 271, 2856–2862 (1996).

82. A. R. van Rompay, M. Johansson and A. Karlsson, Phosphorylation of deoxycytidine analog monophosphates by UMP-CMP kinase: Molecular characterization of the human enzyme. *Mol. Pharmacol.* 56, 562–569 (1999).

83. S. W. Hall and H. Kühn, Purification and properties of guanylate kinase from bovine retinas and rod outer segments. *Eur. J. Biochem.* 161, 551–556 (1986).

84. W. H. Miller and R. L. Miller, Phosphorylation of acyclovir (acycloguanosine) monophosphate by GMP kinase. *J. Biol. Chem.* 255, 7204–7207 (1980).

85. Y. Li, Y. Zhang and H. Yan, Kinetic and thermodynamic characterizations of yeast guanylate kinase. *J. Biol. Chem.* 271, 28038–28044 (1996).

86. L. Serina, C. Blondin, K. E., O. Sismeiro, A. Danchin, H. Sakamoto, A.-M. Gilles and O. Bârzu, *Escherichia coli* UMP-kinase, a member of the aspartokinase family, is a hexamer regulated by guanine nucleotides and UTP. *Biochemistry* 34, 5066–5074 (1995).

87. T. W. Traut, The functions and consensus motifs of nine types of peptide segments that form different types of nucleotide-binding sites. *Eur. J. Biochem.* 222, 9–19 (1994).

88. N. N. Suzuki, K. Koizumi, M. Fukushima, A. Matsuda and F. Inagaki, Structural basis for the specificity, catalysis, and regulation of human uridine–cytidine kinase. *Structure* 12, 751–764 (2004).

89. K. Umezu, T. Amaya, A. Yoshimoto and K. Tomita, Purification and properties of orotidine-5'-phosphate pyrophosphorylase and orotidine-5'-phosphate decarboxylase from baker's yeast. *J. Biochem. (Tokyo)* 70, 249–262 (1971).

90. J. Victor, L. B. Greenberg and D. L. Sloan, Studies of the kinetic mechanism of orotate phosphoribosyltransferase from yeast. *J. Biol. Chem.* 254, 2647–2655 (1979).

91. P. Reyes and R. B. Sandquist, Purification of orotate phosphoribosyltransferase and orotidylate decarboxylase by affinity chromatography on Sepharose dye derivatives. *Anal. Biochem.* 88, 522–531 (1978).

92. P. Reyes and M. Guganig, Studies on a pyrimidine phosphoribosyltransferase from murine leukemia P1534J. *J. Biol. Chem.* 250, 5097–5108 (1975).

93. M. B. Bhatia, A. Vinitsky and C. Grubmeyer, Kinetic mechanism of orotate phosphoribosyltransferase from *Salmonella typhimurium. Biochemistry* 29, 10480–10487 (1990).

94. S. R. Krungkrai, B. J. DelFraino, J. A. Smiley, P. Prapunwattana, T. Mitamura, T. Horii and J. Krungkrai, A novel enzyme complex of orotate phosphoribosyltransferase and orotidine 5'-monophosphate decarboxylase in human malaria parasite *Plasmodium falciparum*: Physical association, kinetics, and inhibition characterization. *Biochemistry* 44, 1643–1652 (2005).

95. R. S. Brody and F. H. Westheimer, The purification of orotidine-5'-phosphate decarboxylase from yeast by affinity chromatography. *J. Biol. Chem.* 254, 4238–4244 (1979).

96. U. Strych, S. Wohlfarth and U. K. Winkler, Orotidine-5'-monophosphate decarboxylase from *Pseudomonas aeruginosa* PAO1: Cloning, overexpression, and enzyme characterization. *Curr. Microbiol.* 29, 353–359 (1994).

97. K. Shostak and M. E. Jones, Orotidylate decarboxylase: insights into the catalytic mechanism from substrate specificity studies. *Biochemistry* 31, 12155–12161 (1992).

98. L. R. Livingstone and M. E. Jones, The purification and preliminary characterization of UMP synthase from human placenta. *J. Biol. Chem.* 262, 15726–15733 (1987).

99. P. Harris, J.-C. N. Poulsen, K. F. Jensen and S. Larsen, Structural basis for the catalytic mechanism of a proficient enzyme: orotidine 5'-monophosphate decarboxylase. *Biochemistry* 39, 4217–4224 (2000).

100. T. W. Traut and B. R. S. Temple, The chemistry of the reaction determines the invariant amino acids during the evolution and divergence of orotidine-5'-monophosphate decarboxylase. *J. Biol. Chem.* 275, 28675–28681 (2000).

101. M. M. Matthews, W. Liao, K. L. Kvalnes-Krick and T. W. Traut, β-Alanine synthase: purification and allosteric properties. *Arch. Biochem. Biophys.* **293**, 254–263 (1992).
102. S. N. Thorn, R. G. Daniels, M.-T. M. Auditor and D. Hilvert, Large rate accelerations in antibody catalysis by strategic use of haptenic charge. *Nature* **373**, 228–230 (1995).
103. J. M. Ackermann, S. Kanugula and A. E. Pegg, DNAzyme-mediated silencing of ornithine decarboxylase. *Biochemistry* **44**, 2143–2152 (2005).
104. S. Sawata, T. Shimayama, M. Komiyama, P. K. R. Kumar, S. Nishikawa and K. Taira, Enhancement of the cleavage rates of DNA-armed hammerhead ribozymes by various divalent metal ions. *Nucl. Acids Res.* **21**, 5656–5660 (1993).
105. Q. Xiang, P. Z. Qin, W. J. Michels, K. Freeland and A. M. Pyle, Sequence specificity of a group II intron ribozyme: multiple mechanisms for promoting unusually high discrimination against mismatched targets. *Biochemistry* **37**, 3839–3849 (1998).
106. L. A. Hegg and M. J. Fedor, Kinetics and therrmodynamics of intermolecular catalysis by hairpin ribozymes. *Biochemistry* **34**, 15813–15828 (1995).
107. A. Sievers, M. Beringer, M. V. Rodnina and R. Wolfenden, The ribosome as an entropy trap. *Proc. Natl Acad. Sci. USA* **101**, 7897–7901 (2004).
108. E. H. Ekland, J. W. Szostak and D. P. Bartel, Structurally complex and highly active RNA ligases derived from random RNA sequences. *Science* **269**, 364–370 (1995).
109. F. Noller and T. R. Cech, Aminoacyl esterase activity of the Tetrahymena ribozyme. *Science* **256**, 1420–1425 (1992).
110. M. J. Fedor and O. C. Uhlenbeck, Kinetics of intermolecular cleavage by hammerhead ribozymes. *Biochemistry* **31**, 12042–12054 (1992).
111. A. Gutteridge and J. Thornton, Understanding nature's toolkit. *Trends Biochem. Sci.* **30**, 622–629 (2005).
112. M. J. Fedor and J. R. Williamson, The catalytic diversity of RNAs. *Nat. Rev. Mol. Cell. Biol.* **6**, 399–412 (2005).
113. R. F. Gesteland, T. R. Cech and J. F. Atkins, *The RNA World, Second Edition.* (Cold Spring Harbor Labortory Press, Cold Spring Harbor, 1999).
114. J. E. Barrick and R. R. Breaker, The power of riboswitches. *Sci. Am.* **296**, 50–57 (2007).
115. C. Huber and G. Wächtershäuser, Peptides by activation of amino acids with CO on (Ni,Fe)S surfaces: implications for the origin of life. *Science* **281**, 670–672 (1998).
116. C. Huber and G. Wächtershäuser, α-Hydroxy and α-amino acids under possible Hadean, volcanic origin-of-life conditions. *Science* **314**, 630–632 (2006).
117. N. Ban, P. Nissen, J. Hansen, P. B. Moore and T. A. Steitz, The complete atomic structure of the large ribosomal subunit at 2.4 Å resolution. *Science* **289**, 905–920 (2000).
118. H. Lineweaver and D. Burk, The determination of enzyme dissociation constants. *J. Am. Chem. Soc.* **56**, 658–666 (1934).
119. G. S. Eadie, The inhibition of cholinesterase by physostigmine and prostigmine. *J. Biol. Chem.* **146**, 85–93 (1942).
120. B. H. J. Hofstee, On the evaluation of the constants V_m and K_M in enzyme reactions. *Science* **116**, 329–331 (1952).
121. G. Scatchard, The attraction of proteins for small molecules and ions. *Ann. N Y Acad. Sci.* **51**, 660–672 (1949).
122. C. S. Hanes, Studies on plant amylases. I. The effect of starch concentration upon the velocity of hydrolysis by the amylase of germinated barley. *Biochem. J.* **26**, 1406–1421 (1932).
123. A. V. Hill, The possible effects of the aggregation of the molecules of haemoglobin on its dissociation curve. *J. Physiol. (Lond).* **40**, iv–vii (1910).
124. G. E. Briggs and J. B. S. Haldane, A note on the kinetics of enzyme action. *Biochem. J.* **19**, 338–339 (1925).
125. J. H. Schwartz and F. Lipmann, Phosphate incorporation into alkaline phosphatase of *E. coli. Proc. Natl. Acad. Sci. USA*, **47**, 1996–2005 (1961).
126. P. English, W. Min, A. M. van Oijen, K. T. Lee, G. Luo, H. Sun, B. J. Cherayil, S. C. Kou and X. S. Xie, Ever-fluctuating single enzyme molecules: Michaelis–Menten equation revisited. *Nat. Chem. Biol.* **2**, 168–175 (2006).
127. H. Pan, J. C. Lee and V. J. Hilser, Binding sites in *Escherichia coli* dihydrofolate reductse communicate by modulating the conformational ensemble. *Proc. Natl Acad. Sci. USA* **97**, 12021–12025 (2000).
128. J. A. Hardy, J. Lam, J. T. Nguyen, T. O'Brien and J. A. Wells, Discovery of an allosteric site in the caspases. *Proc. Natl Acad. Sci. USA* **101**, 12461–12466 (2004).

129. S. W. Wright, A. A. Carlo, M. D. Carty, D. E. Danley, D. L. Hageman, G. A. Karam, C. B. Levy, M. N. Mansour, A. M. Mathiowetz, L. D. McClure, N. B. Nestor, R. K. McPherson, J. Pandit, L. R. Pustilnik, G. K. Schulte, W. C. Soeller, J. L. Treadway, I.-K. Wang, and P. H. Bauer, Anilinoquinazoline inhibitors of fructose 1,6-bisphosphatase bind at a novel allosteric site: synthesis, in vitro characterization, and X-ray crystallography. *J. Med. Chem.* **45**, 3865–3877 (2002).

130. J. Grimsby, R. Sarabu, W. L. Corbett, N.-E. Haynes, F. T. Bizzarro, J. W. Coffey, K. R. Guertin, D. W. Hilliard, R. F. Kester, M. P. E., L. Marcus, L. Qi, C. L. Spence, J. Tengi, M. A. Magnuson, C. A. Chu, M. T. Dvorozniak, F. M. Matschinsky, and J. F. Grippo, Allosteric activators of glucokinase: potential role in diabetes therapy. *Science* **301**, 370–373 (2003).

131. W. H. Martin, D. J. Hoover, S. J. Armento, I. A. Stock, R. K. McPherson, D. E. Danley, R. W. Stevenson, E. J. Barrett and J. L. Treadway, Discovery of a human glycogen phosphorylase inhibitor that lowers blood glucose in vivo. *Proc. Natl Acad. Sci. USA* **95**, 1776–1781 (1998).

132. V. L. Rath, M. Ammirati, D. E. Danley, J. L. Ekstrom, E. M. Gibbs, T. T. Hynes, A. M. Mathiowetz, R. K. McPherson, T. V. Olson, J. L. Treadway, and D. J. Hoover, Human liver glycogen phosphorylase inhibitors bind at a new allosteric site. *Chem. Biol.* **7**, 677–682 (2000).

133. C. Pargelis, L. Tong, L. Churchill, P. M. Cirillo, E. R. Hickey, N. Moss, S. Pav and J. Regan, Inhibition of p38 MAP kinase by utilizing a novel allosteric binding site. *Nature Struct. Biol.* **9**, 268–272 (2002).

134. C. Wiesmann, K. J. Barr, J. Zhu, D. A. Erlanson, W. Shen, B. J. Fahr, M. Zhong, L. Taylor, M. Randal, R. S. McDowell, and S. K. Hansen, Allosteric inhibition of protein tryrosine phosphatase 1B. *Nature Struct. Biol.* **11**, 730–737 (2004).

135. H. K. Jensen, N. Mikkelsen and J. Neuhard, Recombinant uracil phosphoribosyltransferase from the thermophile *Bacillus caldolyticus*: expression, purification, and partial characterization. *Protein Expr. Purif.* **10**, 356–364 (1997).

136. K. F. Jensen and B. Mygind, Different oligomeric states are involved in the allosteric behaviour of uracil phosphoribosyltransferase from *Escherichia coli*. *Eur. J. Biochem.* **204**, 637–645 (1996).

137. M. A. Schumacher, C. J. Bashor, M. H. Song, K. Otsu, S. Zhu, R. J. Parry, B. Ullman and R. G. Brennan, The structural mechanism of GTP stabilized oligomerization and catalytic activation of the *Toxoplasma gondii* uracil phosphoribosyltransferase. *Proc. Natl. Acad. Sci. USA* **99**, 78–83 (2002).

138. H. K. Jensen, S. Arent, S. Larsen and L. Schack, Allosteric properties of the GTP activated and CTP inhibited uracil phosphoribosyltransferase from the thermoacidophilic archaeon *Sulfolobus solfataricus*. *FEBS Journal* **272**, 1440–1453 (2005).

139. M. A. Schumacher, D. Carter, D. M. Scott, D. S. Roos, B. Ullman and R. G. Brennan, Crystal structures of the *Toxoplasma gondii* uracil phosphoribosyltransferase reveal the atomic basis of pyrimidine discrimination and prodrug binding. *EMBO J.* **17**, 3219–3232 (1998).

140. A. Levitzki, W. B. Stallcup and D. E. Koshland Jr., Half-of-the-sites reactivity and the conformational states of cytidine triphosphate synthetase. *Biochemistry* **10** 3371–3378 (1971).

141. M. Mandal, M. Lee, J. E. Barrick, Z. Weinberg, G. M. Emilsson, W. L. Ruzzo and R. R. Breaker, A glycine-dependent riboswitch that uses cooperative binding to control gene expression. *Science* **306** 275–279 (2004).

142. D. E. Koshland Jr., The era of pathway quantification. *Science* **280**, 852–853 (1998).

143. A. Levitzki and D. E. Koshland Jr., Negative cooperativity in regulatory enzymes. *Proc. Natl Acad. Sci. USA* **62**, 1121–1128 (1969).

144. W. B. Stallcup and D. E. Koshland Jr., Half-of-the-sites reactivity and negative cooperativity: The case of yeast glyceraldehyde 3-phosphate dehydrogenase. *J. Mol. Biol.* **80**, 41–62 (1973).

145. P. A. Ropp and T. W. Traut, Purine nucleoside phosphorylase: allosteric regulation of a dissociating enzyme. *J. Biol. Chem.* **266**, 7682–7687 (1991).

146. D. E. Koshland Jr., Correlation of structure and function in enzyme action. *Science* **142**, 1533–1541 (1963).

147. G. G. Hammes, Multiple conformational changes in enzyme catalysis. *Biochemistry* **41**, 8221–8228 (2002).

148. A. Gutteridge and J. Thornton, Conformational changes observed in enzyme crystal structures upon substrate binding. *J. Mol. Biol.* **346**, 21–28 (2005).

149. P. B. Chock and E. R. Stadtman, Superiority of interconvertible enzyme cascades in metabolic regulation: analysis of multicyclic systems. *Proc. Natl Acad. Sci. USA* **74**, 2766–2770 (1977).

150. P. B. Chock, S. G. Rhee and E. R. Stadtman, Interconvertible enzyme cascades in cellular regulation. *Annu. Rev. Biochem.* **49**, 813–843 (1980).

151. E. R. Stadtman and P. B. Chock, Interconvertible enzyme cascades in metabolic regulation. *Curr. Top. Cell. Regul.* **13**, 53–95 (1978).

152. J. E. Ferrel Jr. and E. M. Machleder, The biochemical basis of an all-or-none cell fate switch in *Xenopus* oocytes. *Science* **280**, 895–898 (1998).

153. M. Kennedy, M. Droser, L. M. Mayer, D. Pevear and D. Mrofka, Late precambrian oxygenation: inception of the clay mineral factory. *Science* **311**, 1446–1449 (2006).

154. J. Raymond and D. Segrè, The effect of oxygen on biochemical networks and the evolution of complex life. *Science* **311**, 1764–1767 (2006).

155. R. E. Dickerson and I. Geis, *Hemoglobin.* (Benjamin/Cummings Publishing Co., Menlo Park, CA, 1983).

156. W. E. Royer Jr., H. Zhu, T. A. Gorr, J. F. Flores and J. E. Knapp, Allosteric hemoglobin assembly: diversity and similarity. *J. Biol. Chem.* **280**, 27477–27450 (2005).

157. L. Giangiacomo, M. Mattu, A. Arcovito, G. Bellenchi, M. Bolognesi, P. Ascenzi and A. Boffi, Monomer-dimer equilibrium and oxygen binding properties of ferrous *Vitreoscilla* hemoglobin. *Biochemistry* **40**, 9311–9316 (2001).

158. W. E. Royer Jr., J. E. Knapp, K. Strand and H. A. Heaslet, Cooperative hemoglobins: conserved fold, diverse quaternary assemblies and allosteric mechanisms. *Trends Biochem. Sci.* **26**, 297–304 (2001).

159. C. Bohr, K. Hasselbalch and A. Krogh, Ueber einen in biologischer Beziehung wichtigen Einfluss, den die Kohlsäurespannung des Blutes auf dessen Sauerstoffbindung übt. *Skand. Arch. Physiol.* **16**, 402–412 (1904).

160. M. F. Perutz, H. Muirhead, J. M. Cox and L. C. G. Goaman, Three-dimensional Fourier synthesis of horse oxyhaemoglobin at 2.8 Å resolution: the atomic model. *Nature* **219**, 131–139 (1968).

161. J. Flatley, J. Brrett, S. T. Pullan, M. N. Hughes, J. Green and R. K. Poole, Transcriptional responses of *Escherichia coli* to S-nitrosoglutathione under defined chemostat conditions reveal major changes in methionine biosynthesis. *J. Biol. Chem.* **280**, 10065–10072 (2005).

162. C. A. Appleby, J. H. Bradbury, R. J. Morris, B. A. Wittenberg, J. B. Wittenberg and P. E. Wright, Leghemoglobin. Kinetic, nuclear magnetic resonance, and optical studies of pH dependence of oxygen and carbon monoxide binding. *J. Biol. Chem.* **258**, 2254–2259 (1983).

163. R. Oshino, N. Oshino and B. Chance, Studies on yeast hemoglobin. *Eur. J. Biochem.* **35**, 23–33 (1973).

164. T. Yonetani, S. Parks, A. Tsuneshige, K. Imai and K. Kanaori, Global allostery model of hemoglobin. Modulation of O_2 affinity, cooperativity, and Bohr effect by heterotropic allosteric effectors. *J. Biol. Chem.* **277**, 34508–34520 (2002).

165. S. G. Gross and P. Lane, Physiological reactions of nitric oxide and hemoglobin: a radical rethink. *Proc. Natl Acad. Sci. USA* **96**, 9967–9969 (1999).

166. M. Brunori, A. Giuffrè, K. Nienhaus, G. U. Nienhaus, F. M. Scandurra and B. Vallone, Neuroglobin, nitric oxide, and oxygen: functional pathways and conformational changes. *Proc. Natl Acad. Sci. USA* **102**, 8483–8488 (2005).

167. E. Di Cera, M. L. Doyle, M. S. Morgan, R. De Cristofaro, R. Landolfi, B. Bizzi, M. Castagnola and S. J. Gill, Carbon monoxide and oxygen binding to human hemoglobin F_0. *Biochemistry* **28**, 2631–2638 (1989).

168. J. T. Trent, III and M. S. Hargrove, A ubiquitously expressed human hexacoordinate hemoglobin. *J. Biol. Chem.* **277**, 19538–19545 (2002).

169. T. Burmester, B. Weich, S. Reinhardt and T. Hankeln, A vertebrate globin expressed in the brain. *Nature* **407**, 520–523 (2000).

170. S. Dewilde, L. Kiger, T. Burmester, T. Hankeln, V. Baudin-Creuza, T. Aerts, M. C. Marden, R. Caubergs and L. Moens, Biochemical characterization and ligand binding properties of Neuroglobin, a novel member of the globin family. *J. Biol. Chem.* **276**, 38949–38955 (2001).

171. J. S. Stamler, L. Jia, J. P. Eu, T. J. McMahon, I. T. Demchenko, J. Bonaventura, K. Gernert and C. A. Piantadosi, Blood flow regulation by S-nitrosohemoglobin in the physiological oxygen gradient. *Science* **276** 2034–2037 (1997).

172. W. Chen, A. Dumoulin, X. Li, J. C. Padovan, B. T. Chait, R. Buonopane, O. S. Platt, L. R. Manning and J. M. Manning, Transposing sequences between fetal and adult hemoglobins indicates which subunits and regulatory molecule interfaces are functionally related. *Biochemistry* **39**, 3774–3781 (2000).

173. A. Fago, C. Hundahl, S. Dewilde, K. Gilany, L. Moens and R. E. Weber, Allosteric regulation and temperature dependence of oxygen binding in human neuroglobin and cytoglobin. *J. Biol. Chem.* **279**, 4417–4426 (2004).

174. Y. Sun, K. Jin, A. Peel, X. O. Mao, L. Xie and D. A. Greenberg, Neuroglobin protects the brain from experimental stroke *in vivo. Proc. Natl Acad. Sci. USA* **100**, 3497–3500 (2003).

175. A. Fago, C. Hundahl, H. Malte and R. E. Weber, Functional properties of neuroglobin and cytoglobin. Insights into the ancestral physiological roles. *Life* **56**, 689–696 (2004).

176. A. Pesce, M. Bolognesi, A. Bocedi, P. Ascenzi, S. Dewilde, L. Moens, T. Hankeln and T. Burmester, Neuroglobin and cytoglobin. Fresh blood for the vertebrate globin family. *EMBO Reports* **3**, 1146–1151 (2002).

177. K. Imai, Analyses of oxygen equilibria of native and chemically modified human adult hemoglobin on the basis of Adair's stepwise oxygenation theory and the allosteric model of Monod, Wyman, and Changeux. *Biochemistry* **12**, 798–808 (1973).

178. M. F. Perutz, A. J. Wilkinson, M. Paoli and G. G. Dodson, The stereochemical mechanism of the cooperative effects in hemoglobin revisited. *Annu. Rev. Biophys. Biomol. Struct.* **27**, 1–34 (1998).

179. J. V. Kilmartin and L. Rossi-Bernardi, Interaction of hemoglobin with hydrogen ions, carbon dioxide, and organic phosphates. *Physiol. Rev.* **53**, 836–890 (1973).

180. K. Imai and T. Yonetani, Thermodynamical studies of oxygen equilibrium of hemoglobin. *J. Biol. Chem.* **250**, 7093–7098 (1975).

181. C. Rivetti, A. Mozzarelli, G. L. Rossi, E. R. Henry and W. A. Eaton, Oxygen binding by single crystals of hemoglobin. *Biochemistry* **32**, 2888–2906 (1993).

182. R. E. Benesch and R. Benesch, The mechanism of interaction or red cell organic phosphates with hemoglobin. *Adv. Protein Chem.* **28**, 211–237 (1974).

183. J. S. Olson, Binding of inositol hexaphosphate to human methemoglobin. *J. Biol. Chem.* **251**, 447–458 (1976).

184. G. S. Adair, The hemoglobin system. VI The oxygen dissociation curve of hemoglobin. *J. Biol. Chem.* **63**, 529–545 (1925).

185. N. Shibayama and S. Saigo, Fixation of the quaternary structures of human adult hemoglobin by encapsulation in transparent porous silica gels. *J. Mol. Biol.* **251**, 203–209 (1995).

186. N. B. Livanova, N. A. Chebotareva, T. B. Eronina and B. I. Kurganov, Pyridoxal 5'-phosphate as a catalytic and conformational cofactor of muscle glycogen phosphorylase *b*. *Biochemistry (Moscow)* **67**, 1317–1327 (2002).

187. P. J. Keller and G. T. Cori, Enzymic conversion of phosphorylase *a* to phosphorylase *b*. *Biochim. Biophys. Acta* **12**, 235–238 (1953).

188. R. J. Fletterick and N. B. Madsen, The structures and related functions of phosphorylase *a*. *Annu. Rev. Biochem.* **49**, 31–61 (1980).

189. A. B. Kent, E. G. Krebs and E. H. Fischer, Properties of crystalline phosphorylase *b*. *J. Biol. Chem.* **232**, 549–558 (1958).

190. J. L. Hedrick, S. Shaltiel and E. H. Fischer, On the role of pyridoxal 5'-phosphate in phosphorylase. 3. Physicochemical properties and reconstitution of apophosphorylase *b*. *Biochemistry* **5**, 2117–2125 (1966).

191. P. J. Kasvinsky and W. L. Meyer, The effect of pH and temperature on the kinetics of native and altered glycogen phosphorylase. *Arch. Biochem. Biophys.* **181**, 616–631 (1977).

192. N. B. Madsen, S. Schechosky and R. J. Fletterick, Site–site interactions in glycogen phosphorylase *b* probed by ligands specific for each site. *Biochemistry* **22**, 4460–4465 (1983).

193. G. T. Cori and C. F. Cori, The kinetics of the enzymatic synthesis of glycogen from glucose-1-phosphate. *J. Biol. Chem.* **135**, 733–756 (1940).

194. S. R. Sprang, K. R. Acharya, E. J. Goldsmith, D. I. Stuart, K. M. Varvill, R. J. Fletterick and N. B. Madsen, Protein phosphorylation: structural change between glycogen phosphorylase *b* and *a*. *Nature* **336**, 215–221 (1988).

195. W. J. Black and J. H. Wang, Studies on the allosteric activation of glycogen phosphorylase *b* by nucleotides. *J. Biol. Chem.* **243**, 5892–5898 (1968).

196. V. L. Rath, C. B. Newgard, S. R. Sprang, E. J. Goldsmith and R. J. Fletterick, Modeling of the biochemical differences between rabbit muscle and human liver phosphorylase. *Prot. Str. Fn. Gen.* **2**, 225–235 (1987).

197. L. N. Johnson, Glycogen phosphorylase: control by phosphorylation and allosteric effectors. *FASEB J.* **6**, 2274–2282 (1992).

198. D. Palm, H. Goerl and K. J. Burger, Evolution of catalytic and regulatory sites in phosphorylases. *Nature* **313**, 500–502 (1985).

199. C. B. Newgard, P. K. Hwang and R. J. Fletterick, The family of glycogen phosphorylases: structure and function. *CRC Crit. Rev. Biochem. Mol. Biol.* **24**, 69–99 (1989).

200. D. Guénard, M. Morange and H. Buc, Comparative study of the effect of 5'-AMP and its analogs on rabbit glycogen phosphorylase *b* isoenzymes. *Eur. J. Biochem.* **76**, 447–452 (1977).

201. E. Mertens, Pyrophosphate-dependent phosphofructokinase, an anaerobic glycolytic enzyme? *FEBS Lett.* **285**, 1–5 (1991).

202. R. S. Ronimus and H. W. Morgan, The biochemical properties and phylogenies of phosphofructokinases from extremophiles. *Extremophiles* **5**, 357–373 (2001).

203. D. Blangy, H. Buc and J. Monod, Kinetics of the allosteric interactions of phosphofructokinase from *Escherichia coli. J. Mol. Biol.* **31**, 13–35 (1968).

204. J. Heinisch, R. G. Ritzel, R. C. von Borstel, A. Aguilera, R. Rodicio and F. K. Zimmermann, The phosphofructokinase genes of yeast evolved from two duplication events. *Gene* **78**, 309–321 (1989).

205. O. H. Martinez-Costa, A. M. Estévez, V. Sánchez and J. J. Aragón, Purification and properties of phosphofructokinase from *Dictyostelium discoidium. Eur. J. Biochem.* **226**, 1007–1017 (1994).

206. K. Uyeda, E. Furuya and L. J. Luby, The effect of natural and synthetic D-fructose 2,6-bisphosphate on the regulatory kinetic properties of liver and muscle phosphofructokinases. *J. Biol. Chem.* **256**, 8394–8399 (1981).

207. J. E. Tuininga, C. H. Verhees, J. van der Oost, S. W. M. Kengen, A. J. M. Stams and W. M. de Vos, Molecular and biochemical characterization of the ADP-dependent phosphofructokinase from the hyperthermophilic *Pyrococcus furiosus. J. Biol. Chem.* **274**, 1023–1028 (1999).

208. T. Enomoto, K. Miyatake and S. Kitaoka, Purification and immunological properties of fructose-2,6-bisphosphate-sensitive pyrophosphate: D-fructose 6-phosphate 1-phosphotransferase from the protist *Euglena gracilis. Comp. Biochem. Physiol.* **90B**, 897–902 (1988).

209. K. Miyatake, T. Enomoto and S. Kitaoka, Detection and subcellular distribution of pyrophosphate: D-fructose 6-phosphate phosphotransferase (PFP) in *Euglena gracilis. Agric. Biol. Chem.* **48**, 2857–2859 (1984).

210. D. Fell, *Understanding the Control of Metabolism* (Portland Press, London, 1997).

211. A. M. C. R. Alves, G. J. W. Euverink, H. Santos and L. Dijkhuizen, Different physiological roles of ATP- and PP$_i$-dependent phosphofructokinase isoenzymes in the methylotropic actinomycete *Amycolatopsis methanolica. J. Bacteriol.* **183**, 7231–7240 (2001).

212. T. Schirmer and P. R. Evans, Structural basis of the allosteric behaviour of phosphofructokinase. *Nature* **343**, 140–145 (1990).

213. F. T.-K. Lau and A. R. Fersht, Dissection of the effector-binding site and complementation studies of *Escherichia coli* phosphofructokinase using site-directed mutagenesis. *Biochemistry* **28**, 6841–6847 (1989).

214. S. Kitajima, R. Sakakibara and K. Uyeda, Significance of phosphorylation of phosphofructokinase. *J. Biol. Chem.* **258**, 13292–13298 (1983).

215. R. Sakakibara and K. Uyeda, Differences in the allosteric properties of pure low and high phosphate forms of phosphofructokinase from rat liver. *J. Biol. Chem.* **258**, 8656–8662 (1983).

216. M. Bañuelos, C. Gancedo and J. M. Gancedo, Activation by phosphate of yeast phosphofructokinase. *J. Biol. Chem.* **252**, 6394–6398 (1977).

217. G. A. Dunaway, T. P. Kasten, T. Sebo and R. Trapp, Analysis of the phosphofructokinase subunits and isoenzymes in human tissues. *Biochem. J.* **251**, 677–683 (1988).

218. J. L. Johnson and G. D. Reinhart, Failure of a two-state model to describe the influence of phospho(*enol*)pyruvate on phosphofructokinase from *Escherichia coli. Biochemistry* **36**, 12814–12822 (1997).

219. A. Jordan and P. Reichard, Ribonucleotide reductases. *Annu. Rev. Biochem.* **67**, 71–98 (1998).

220. P. Nordlund and P. Reichard, Ribonucleotide reductases. *Annu. Rev. Biochem.* **75**, 681–706 (2006).

221. U. Uhlin and H. Eklund, Structure of ribonucleotide reductase protein R1. *Nature* **370**, 533–539 (1994).

222. D. T. Logan, J. Andersson, B.-M. Sjöberg and P. Nordlund, A glycyl radical site in the crystal structure of a class III ribonucleotide reductase. *Science* **283**, 1499–1504 (1999).

223. P. Reichard, Ribonucleotide reductases: the evolution of allosteric regulation. *Arch. Biochem. Biophys.* **397**, 149–155 (2002).

224. H. Eklund and M. Fontecave, Glycyl radical enzymes: a conservative structural basis for radicals. *Structure* **7**, R257–R262 (1999).

225. J. Stubbe and P. Riggs-Gelasco, Harnessing free radicals: formation and function of the tyrosyl radical in ribonucleotide reductase. *Trends Biochem. Sci.* **23**, 438–443 (1998).

226. N. C. Brown and P. Reichard, Role of effector binding in allosteric control of ribonucleotide diphosphate reductase. *J. Mol. Biol.* **46**, 39–55 (1969).

227. N. C. Brown and P. Reichard, Ribonucleotide diphosphate reductase. Formation of active and inactive complexes of proteins B1 and B2. *J. Mol. Biol.* **46**, 25–38 (1969).

228. S. Eriksson, L. Thelander and M. Åkerman, Allosteric regulation of calf thymus ribonucleoside diphosphate reductase. *Biochemistry* **18**, 2948–2952 (1979).

229. S. Eriksson, A. Gräslund, S. Skog, L. Thelander and B. Tribukait, Cell cycle-dependent regulation of mammalian ribonucleotide reductase. *J. Biol. Chem.* **259**, 11695–11700 (1984).

230. J. Shao, B. Zhou, L. Zhu, W. Qiu, Y.-C. Yuan, B. Xi and Y. Yen, In vitro characterization of enzymatic properties and inhibition of the p53R2 subunit of human ribonucleotide reductase. *Cancer Res.* **64**, 1–6 (2004).

231. W. Qiu, B. Zhou, D. Darwish, J. Shao and Y. Yen, Characterization of enzymatic properties of human ribonucleotide reductase holoenzyme reconstituted in vitro from hRRM1, hRRM2, and p53R2 subunits. *Biochem. Biophys. Res. Commun.* **340**, 428–434 (2006).

232. Y. Yen, B. Chu, C. Yen, J. Shih and B. Zhou, Enzymatic property analysis of the p53R2 subunit of human ribonucleotide reductase. *Adv. Enzyme Regul.* **46**, 235–247 (2006).

233. V. M. Monnier, Intervention against the Maillard reaction in vivo. *Arch. Biochem. Biophys.* **419**, 1–15 (2003).

234. A. E. Aleshin, M. Malfois, X. Liu, C. S. Kim, H. J. Fromm, R. B. Honzatko, M. H. J. Koch and D. L. Svergun, Nonaggregating mutant of recombinant human hexokinase I exhibits wild-type kinetics and rod-like conformations in solution. *Biochemistry* **38**, 8359–8366 (1999).

235. F. Palma, D. Agostini, P. Mason, M. Dachà, G. Piccoli, B. Biagiarelli, M. Fiorani and V. Stocchi, Purification and characterization of the carboxyl-domain of human hexokinase type III expressed as fusion protein. *Mol. Cell. Biochem.* **155**, 23–29 (1996).

236. A. L. Gloyn, S. Odili, D. Zelent, C. W. Buettger, H. A. J. Castleden, A. M. Steele, A. Stride, C. Shiota, M. A. Magnusson, R. Lorini, G. d'Annunzio, C. A. Stanley, J. Kwagh, E. Van Schaftingen, M. Veiga-da-Cunha, F. Barbetti, P. Dunten, Y. Han, J. Grimsby, R. Taub, S. Ellard, A. T. Hattersley, and F. M. Matschinsky, Insights into the structure and regulation of glucokinase from a novel mutation (V62M), which causes maturity-onset diabetes of the young. *J. Biol. Chem.* **280**, 14105–14113 (2005).

237. M. Magnani, V. Stocchi, N. Serafini, E. Piatti, M. Dachà and G. Fornaini, Pig red blood cell hexokinase: regulatory characteristics and possible physiological role. *Arch. Biochem. Biophys.* **226**, 377–387 (1983).

238. E. Van Schaftingen, Short-term regulation of glucokinase. *Diabetologia* **37**, S43–S47 (1994).

239. V. Stocchi, M. Magnani, G. Novelli, M. Dachà and G. Fornaini, Pig red blood cell hexokinase: evidence for the presence of hexokinase types II and III, and their purification and characterization. *Arch. Biochem. Biophys.* **226**, 365–376 (1983).

240. M. L. Cárdenas, A. Cornish-Bowden and T. Ureta, Evolution and regulatory role of the hexokinases. *Biochim. Biophys. Acta* **1401**, 242–264 (1998).

241. R. Golbik, M. Naumann, A. Otto, E.-C. Müller, R. Behlke, R. Reuter, G. Hübner and T. H. Kriegel, Regulation of phosphotransferase activity of hexokinase 3 from *Saccharomyces cerevisiae* by modification of serine-14. *Biochemistry* **40**, 1083–1090 (2001).

242. P. B. Iynedjian, Mammalian glucokinase and its gene. *Biochem. J.* **293**, 1–13 (1993).

243. T.-Y. Fang, O. Alechina, A. E. Aleshin, H. J. Fromm and R. B. Honzatko, Identification of a phosphate regulatory site and a low affinity binding site for glucose-6-phosphate in the N-terminal half of human brain hexokinase. *J. Biol. Chem.* **273**, 19548–19553 (1998).

244. T. K. White and J. E. Wilson, Isolation and characterization of the discrete N- and C-terminal halves of rat brain hexokinase: retention of full catalytic activity in the isolated C-terminal half. *Arch. Biochem. Biophys.* **274**, 375–393 (1989).

245. A. E. Aleshin, C. Kirby, X. Liu, G. B. Bourenkov, H. D. Bartunik, H. J. Fromm and R. B. Honzatko, Crystal structures of mutant monomeric hexokinase I reveal multiple ADP binding sites and conformational changes relevant to allosteric regulation. *J. Mol. Biol.* **296**, 1001–1015 (2000).

246. J. E. Wilson, Isozymes of mammalian hexokinase: structure, subcellular localization and metabolic function. *J. Exp. Biol.* **206**, 2049–2057 (2003).

247. H. BeltrandelRio and J. E. Wilson, Hexokinase of rat brain mitochondria: relative importance of adenylate kinase and oxidative phosphorylation as sources of substrate ATP, and interaction with intramitochondrial compartments of ATP and ADP. *Arch. Biochem. Biophys.* **286**, 183–184 (1991).

248. M. J. Holroyde, M. B. Allen, A. C. Storer, A. S. Warsy, J. M. E. Chesher, J. P. Trayer, A. Cornish-Bowden and D. G. Walker, The purification in high yield and characterization of rat hepatic glucokinase. *Biochem. J.* **153**, 163–373 (1976).

249. M. L. Cárdenas, E. Rabajille and H. Niemeyer, Maintenance of the monomeric structure of glucokinase under reacting conditions. *Arch. Biochem. Biophys.* **190**, 142–148 (1978).

250. M. S. Palma, A. M. Teno and A. Rossi, Acid phosphatase from maize scutellum: negative cooperativity suppression by glucose. *Phytochemistry* **22**, 1899–1901 (1983).
251. E. J. Walker, G. B. Ralston and I. G. Darvey, An allosteric model for ribonuclease. *Biochem. J.* **147**, 425–433 (1975).
252. E. J. Walker, G. B. Ralston and I. G. Darvey, Further evidence for an allosteric model for ribonuclease. *Biochem. J.* **153**, 329–337 (1976).
253. W. S. Beck, Regulation of cobamide-dependent ribonucleotide reductase by allosteric effectors and divalent cations. *J. Biol. Chem.* **242**, 3148–3158 (1967).
254. D. Panagou, M. D. Orr, J. R. Dunstone and R. L. Blakley, Cobamides and ribonucleotide reduction. IX. Monomeric, allosteric enzyme with a single polypeptide chain. Ribonucleotide reductase of *Lactobacillus leichmannii*. *Biochemistry* **11**, 2378–2388 (1972).
255. M. A. Moukil and E. Van Schaftingen, Analysis of the cooperativity of human β-cell glucokinase through the stimulatory effect of glucose on fructose phosphorylation. *J. Biol. Chem.* **2001**, 3872–3878 (2001).
256. G. R. Ainslie Jr., J. P. Shill and K. E. Neet, Transients and cooperativity. A slow transition model for relating transients and cooperative kinetics of enzymes. *J. Biol. Chem.* **247**, 7088–7096 (1972).
257. K. E. Neet, Cooperativity in enzyme function: equilibrium and kinetic aspects, in: *Contemporary Enzyme Kinetics and Mechanism*, D.L. Purich, Editor. (Academic Press, New York, 1983) pp. 267–320.
258. K. E. Neet, R. P. Keenan and P. S. Tippett, Observation of a kinetic slow transition in monomeric glucokinase. *Biochemistry* **29**, 770–777 (1990).
259. H. Buc, Enzyme memory. Effect of glucose 6-phosphate and temperature on the molecular transition of wheat-germ hexokinase LI. *Eur. J. Biochem.* **97**, 573–583 (1977).
260. M. L. Cárdenas, E. Rabajille and H. Niemeyer, Suppression of kinetic cooperativity of hexokinase D (glucokinase) by competitive inhibitors. *Eur. J. Biochem.* **145**, 163–171 (1984).
261. A. Cornish-Bowden, The effect of natural selection on enzyme catalysis. *J. Mol. Biol.* **101**, 1–9 (1976).
262. A. Cornish-Bowden and M. L. Cárdenas, Co-operativity in monomeric enzymes. *J. Theor. Biol.* **124**, 1–23 (1987).
263. K. Kamata, M. Mitsuya and Y. Nagat, Structural basis for allosteric regulation of the monomeric allosteric enzyme human glucokinase. *Structure* **12**, 429–438 (2004).
264. A. E. Aleshin, C. Zeng, H. D. Bartunik, H. J. Fromm and R. B. Honzatko, Regulation of hexokinase I: crystal structure of recombinant human brain hexokinase complexed with glucose and phosphate. *J. Mol. Biol.* **282**, 345–357 (1998).
265. V. V. Heredia, J. Thomson, D. Nettleton and S. Sun, Glucose-induced conformational changes in glucokinase mediate allosteric regulation: transient kinetic analysis. *Biochemistry* **45**, 7553–7562 (2006).
266. Y. B. Kim, S. S. Kalinowski and J. Marcinkeviciene, A pre-steady state analysis of ligand binding to human glucokinase: evidence for a preexisting equilibrium. *Biochemistry* **46**, 1423–1431 (2007).
267. H. Niemeyer, M. L. Cárdenas, E. Rabajille, T. Ureta, L. Clark-Turri and J. Peñaranda, Sigmoidal kinetics of glucokinase. *Enzyme* **20**, 321–333 (1975).
268. A. C. Storer and A. Cornish-Bowden, Kinetic evidence for a 'mnemonical' mechanism for rat liver glucokinase. *Biochem. J.* **159**, 7–14 (1978).
269. E. Van Schaftingen, A protein from rat liver confers to glucokinase the property of being antagonistically regulated by fructose-6-phosphate and fructose-1-phosphate. *Eur. J. Biochem.* **179**, 179–184 (1989).
270. E. Van Schaftingen, M. Veiga-da-Cunha and L. Niculescu, The regulatory protein of glucokinase. *Biochem. Soc. Trans.* **25**, 136–140 (1997).
271. P. S. Tippett and K. E. Neet, Specific inhibition of glucokinase by long chain acyl coenzymes A below the critical micelle concentration. *J. Biol. Chem.* **257**, 12839–12845 (1982).
272. S. Kawai, T. Mukai, S. Mori, B. Mikami and K. Murata, Structures, evolution, and ancestor of glucose kinases in the hexokinase family. *J. Biosci. Bioeng.* **99**, 320–330 (2005).
273. K. K. Arora, C. R. Filburn and J. Pedersen, Structure/function relationships in hexokinase. Site-directed mutational analyses and characterization of overexpressed fragments implicate different functions for the N- and C-terminal halves of the enzyme. *J. Biol. Chem.* **268**, 18259–18266 (1993).
274. C. M. de Cerqueira and J. E. Wilson, Further studies on the coupling of the mitochondrially bound hexokinase to intramitochondrially compartmented ATP, generated by oxidative phosphorylation. *Arch. Biochem. Biophys.* **350**, 109–117 (1998).
275. K. Kaiserova, S. Srivastava, J. D. Hoetker, S. O. Awe, X.-L. Tang, J. Cai and A. Bhatnagar, Redox activation of aldose reductase in the ischemic heart. *J. Biol. Chem.* **281**, 15110–15120 (2006).
276. S. Chacko and E. Eisenberg, Cooperativity of actin-activated ATPase of gizzard heavy meromyosin in the presence of gizzard tropomyosin. *J. Biol. Chem.* **265**, 2105–2110 (1990).

277. N. Nagahara, T. Yoshii, Y. Abe and T. Matsumura, Thioredoxin-dependent enzymatic activation of mercaptopyruvate sulfurtransferase. *J. Biol. Chem.* **282**, 1561–1569 (2007).

278. N. Kumar and U. Varshney, Contrasting effects of single stranded DNA binding protein on the activity of uracil DNA glycosylase from *Escherichia coli* towards different DNA substrates. *Nucl. Acids Res.* **25**, 2336–2343 (1997).

279. L. S. Chesnokova and S. N. Witt, Switches, catapults, and chaperones: steady-state kinetic analysis of HSP70-substrate interactions. *Biochemistry* **44**, 11224–11233 (2005).

280. J. R. Stone and M. A. Marletta, Spectral and kinetic studies on the activation of soluble guanylate cyclase by nitric oxide. *Biochemistry* **35**, 1093–1099 (1996).

281. V. L. Schramm, Kinetic properties of allosteric adenosine monophosphate nucleosidase from *Azotobacter vinelandii. J. Biol. Chem.* **249**, 17291736 (1974).

282. P. Gideon, J. John, M. Frech, A. Lautwein, R. Clark, J. E. Scheffler and A. Wittinghofer, Mutational and kinetic analyses of the GTPase-activating protein (GAP)-p21 interaction: the C-terminal domain of GAP is not sufficient for full activity. *Mol. Cell. Biol.* **12**, 2050–2056 (1992).

283. E. Sugimoto and E. Pizer, The mechanism of end product inhibition of serine biosynthesis. *J. Biol. Chem.* **243**, 2081–2089 (1968).

284. N. F. B. Phillips, M. A. Snoswell, A. Chapman-Smith, D. B. Keech and J. C. Wallace, Isolation of a carboxyphosphate intermediate and the locus of acetyl-CoA action in the pyruvate carboxylase reaction. *Biochemistry* **31**, 9445–9450 (1992).

285. T. W. Traut, All regulatory enzymes are K-type. *(submitted)*, (2007).

286. T. A. J. Haystead and D. G. Hardie, Both insulin and epidermal growth factor stimulate lipogenesis and acetyl-CoA carboxylase activity in isolated adipocytes. *Biochem. J.* **234**, 279–284 (1986).

287. M. Trahey and F. McCormick, A cytoplasmic protein stimulates normal N-ras p21 GTPase, but does not affect oncogenic mutants. *Science* **238**, 542–545 (1987).

288. R. S. Goody, M. Frech and A. Wittinghofer, Affinity of guanine nucleotide binding proteins for their ligands: facts and artefacts. *Trends Biochem. Sci.* **16**, 327–328 (1991).

289. J. John, R. Sohmen, J. Feuerstein, R. Linke, A. Wittinghofer and R. S. Goody, Kinetics of interaction of nucleotides with nucleotide-free H-ras p21. *Biochemistry* **29**, 6058–6065 (1990).

290. M. Fech, T. A. Darden, L. G. Pedersen, C. K. Foley, P. S. Charifson, M. W. Anderson and A. Wittinghofer, Role of glutamine-61 in the hydrolysis of GTP by p21$^{H\text{-}ras}$: an experimental and theoretical study. *Biochemistry* **33**, 3237–3244 (1994).

291. G. D. Plowman, S. Sudarsanam, J. Bingham, D. Whyte and T. Hunter, The protein kinases of Caenorhabditis elegans: a model for signal transduction in multicellular organisms. *Proc. Natl Acad. Sci. USA* **96**, 13603–13610 (1999).

292. T. Fox, M. J. Fitzgibbon, M. A. Fleming, H.-M. Hsiao, C. L. Brummel and M. S.-S. Su, Kinetic mechanism and ATP-binding site reactivity of p38γ MAP kinase. *FEBS Lett.* **461**, 323–328 (1999).

293. B. Stein, M. X. Yang, D. B. Young, R. Janknecht and T. Hunter, p38-2, a novel mitogen-activated protein kinase with distinct properties. *J. Biol. Chem.* **272**, 19509–19517 (1997).

294. M. Avitzour, R. Diskin, B. Raboy, N. Askari, D. Engelberg and O. Livnah, Intrinsically active variants of all human p38 isoforms. *FEBS J.* **274**, 963–975 (2007).

295. J. Raingeaud, S. Gupta, J. S. Rogers, M. Dickens, J. Han, R. J. Ulevitch and R. J. Davis, Pro-inflammatory cytokines and environmental stress causes p38 mitogen-activated protein kinase activation by dual phosphorylation on tyrosine and threonine. *J. Biol. Chem.* **270**, 7420–7426 (1995).

296. C.-Y. F. Huang and J. E. Ferrel Jr., Ultrasensitivity in the mitogen-activated protein kinase cascade. *Proc. Natl Acad. Sci. USA* **93**, 10078–10083 (1996).

297. A. Goldbeter and D. E. Koshland Jr., An amplified sensitivity arising from covalent modification in biological systems. *Proc. Natl Acad. Sci. USA*, **78**, 6840–6844 (1981).

298. D. E. Koshland Jr., A. Goldbeter and J. B. Stock, Amplification and adaptation in regulatory systems. *Science* **217**, 220–225 (1982).

299. D. G. Hardie, D. Carling and M. Carlson, The AMP-activated/SNF1 protein kinase subfamily: metabolic sensors of the eukaryotic cell. *Annu. Rev. Biochem.* **67**, 821–855 (1998).

300. Y. Minokoshi, Y.-B. Kim, O. D. Peroni, L. G. Fryer, C. Müller, D. Carling and B. B. Kahn, Leptin stimulates fatty-acid oxidation by activating AMP-activated protein kinase. *Nature*, **415**, 339–343 (2002).

301. D. Carling and D. G. Hardie, The substrate and sequence specificity of the AMP-activated protein kinase. Phosphorylation of glycogen synthase and phosphorylase kinase. *Biochim. Biophys. Acta* **1012**, 81–86 (1989).

302. J. M. Corton, J. G. Gillespie, S. A. Hawley and D. G. Hardie, 5-Aminoimidazole-4-carboxamide ribonucleoside. A specific method for activating AMP-activated protein kinase in intact cells? *Eur. J. Biochem.* **229**, 558–565 (1995).

303. J. Weekes, S. A. Hawley, J. M. Corton, D. Shugar and D. G. Hardie, Activation of rat liver AMP-activated protein kinase by kinase kinase in a purified, reconstituted system. Effects of AMP and AMP analogues. *Eur. J. Biochem.* **219**, 751–757 (1994).

304. R. Townley and L. Shapiro, Crystal structures of the adenylate sensor from fission yeast AMP-activated protein kinase. *Science* **315**, 1726–1729 (2007).

305. D. Øgreid and S. O. Døskeland, Activation of protein kinase isoenzymes under physiological conditions. Evidence that both types (A and B) of cAMP binding sites are involved in the activation of protein kinase by cAMP and 8-N₃-cAMP. *FEBS Lett.* **150**, 161–166 (1982).

306. D. A. Walsh, J. J. Perkins and E. G. Krebs, An adenosine 3',5'-monophosphate-dependent protein kinase from rabbit skeletal muscle. *J. Biol. Chem.* **243**, 3763–3774 (1968).

307. D. A. Johnson, P. Akamine, E. Radzio-Andxein, Madhusudan and S. S. Taylor, Dynamics of cAMP-dependent protein kinase. *Chem. Rev.* **101**, 2243–2270 (2001).

308. J. Bubis, J. J. Neitzel, L. D. Saraswat and S. S. Taylor, A point mutation abolishes binding of cAMP to site A in the regulatory subunit of cAMP-dependent protein kinase. *J. Biol. Chem.* **263**, 9668–9673 (1988).

309. S. R. Rannels and J. D. Corbin, Two different intrachain cAMP binding sites of cAMP-dependent protein kinases. *J. Biol. Chem.* **255**, 7085–7088 (1980).

INDEX

Printed in the United States of America